Improving Operations AND Long-Term Safety OF THE Waste Isolation Pilot Plant

Final Report

Committee on the Waste Isolation Pilot Plant
Board on Radioactive Waste Management
Division on Earth and Life Studies
National Research Council

NATIONAL ACADEMY PRESS
Washington, D.C.

NOTICE: The project that is the subject of this report was approved by the Governing Board of the National Research Council, whose members are drawn from the councils of the National Academy of Sciences, the National Academy of Engineering, and the Institute of Medicine. The members of the committee responsible for the report were chosen for their special competences and with regard for appropriate balance.

Support for this study was provided by the U.S. Department of Energy under cooperative agreement numbers DE-FC01-94EW54069 and DE-FC01-99EW59049. All opinions, findings, conclusions, and recommendations expressed herein are those of the authors and do not necessarily reflect the views of the Department of Energy.

International Standard Book Number 0-309-07344-8.

Additional copies of this report are available from:
National Academy Press
2101 Constitution Avenue, N.W.
Box 285
Washington, DC 20055
800-624-6242
202-334-3313 (in the Washington Metropolitan Area)
http://www.nap.edu

Cover: The four drawings on the left represent the natural process of salt encapsulating transuranic waste drums in the WIPP repository. From top to bottom, the chronological sequence is 0 years, 10-15 years, 50 years, and 1,000 and more years (Hansen et al., 1997). Reproduced with permission. The image in the center shows a sample of Permian age salt crystals, about 225 million years old, taken from the WIPP excavations. The picture on the right shows typical scenery in proximity to the WIPP repository.

Back cover: Picture of three trucks transporting transuranic waste to the WIPP. *Source:* DOE.

Copyright 2001 by the National Academy of Sciences. All rights reserved.

Printed in the United States of America.

THE NATIONAL ACADEMIES

National Academy of Sciences
National Academy of Engineering
Institute of Medicine
National Research Council

The **National Academy of Sciences** is a private, nonprofit, self-perpetuating society of distinguished scholars engaged in scientific and engineering research, dedicated to the furtherance of science and technology and to their use for the general welfare. Upon the authority of the charter granted to it by the Congress in 1863, the Academy has a mandate that requires it to advise the federal government on scientific and technical matters. Dr. Bruce M. Alberts is president of the National Academy of Sciences.

The **National Academy of Engineering** was established in 1964, under the charter of the National Academy of Sciences, as a parallel organization of outstanding engineers. It is autonomous in its administration and in the selection of its members, sharing with the National Academy of Sciences the responsibility for advising the federal government. The National Academy of Engineering also sponsors engineering programs aimed at meeting national needs, encourages education and research, and recognizes the superior achievements of engineers. Dr. William A. Wulf is president of the National Academy of Engineering.

The **Institute of Medicine** was established in 1970 by the National Academy of Sciences to secure the services of eminent members of appropriate professions in the examination of policy matters pertaining to the health of the public. The Institute acts under the responsibility given to the National Academy of Sciences by its congressional charter to be an adviser to the federal government and, upon its own initiative, to identify issues of medical care, research, and education. Dr. Kenneth I. Shine is president of the Institute of Medicine.

The **National Research Council** was organized by the National Academy of Sciences in 1916 to associate the broad community of science and technology with the Academy's purposes of furthering knowledge and advising the federal government. Functioning in accordance with general policies determined by the Academy, the Council has become the principal operating agency of both the National Academy of Sciences and the National Academy of Engineering in providing services to the government, the public, and the scientific and engineering communities. The Council is administered jointly by both Academies and the Institute of Medicine. Dr. Bruce M. Alberts and Dr. William A. Wulf are chairman and vice chairman, respectively, of the National Research Council.

COMMITTEE ON THE WASTE ISOLATION PILOT PLANT

B. JOHN GARRICK, *Chair,* Garrick Consulting, Laguna Beach, California
MARK D. ABKOWITZ, Vanderbilt University, Nashville, Tennessee
ALFRED W. GRELLA, Grella Consulting, Locust Grove, Virginia
MICHAEL P. HARDY, Agapito Associates, Inc., Grand Junction, Colorado
STANLEY KAPLAN, Bayesian Systems Inc., Rockville, Maryland
HOWARD M. KINGSTON, Duquesne University, Pittsburgh, Pennsylvania
W. JOHN LEE, Texas A&M University, College Station
MILTON LEVENSON, Bechtel International, Inc. (retired), Menlo Park, California
WERNER F. LUTZE, University of New Mexico, Albuquerque
KIMBERLY OGDEN, University of Arizona, Tucson
MARTHA R. SCOTT, Texas A&M University, College Station
JOHN M. SHARP, JR., The University of Texas, Austin
PAUL G. SHEWMON, Ohio State University (retired), Columbus
JAMES E. WATSON, JR., University of North Carolina, Chapel Hill
CHING H. YEW, The University of Texas (retired), Austin

Liaisons

DARLEANE C. HOFFMAN, Lawrence Berkeley National Laboratory, Oakland, California (February 1998 to December 1999)
JAMES O. LECKIE, Stanford University, Stanford, California (January 2000 to December 2000)

Staff

BARBARA PASTINA, Study Director
THOMAS E. KIESS, Study Director (February 1998 to May 2000)
ANGELA R. TAYLOR, Senior Project Assistant

Consultants

LYNDA L. BROTHERS, Sonnenschein Nath & Rosenthal, San Francisco, California
JOHN T. SMITH, Covington & Burlington, Washington, D.C.

BOARD ON RADIOACTIVE WASTE MANAGEMENT

JOHN F. AHEARNE, *Chair*, Sigma Xi and Duke University, Research Triangle Park, North Carolina
CHARLES MCCOMBIE, *Vice-Chair*, Consultant, Gipf-Oberfrick, Switzerland
ROBERT M. BERNERO, U.S. Nuclear Regulatory Commission (retired), Gaithersburg, Maryland
ROBERT J. BUDNITZ, Future Resources Associates, Inc., Berkeley, California
GREGORY R. CHOPPIN, Florida State University, Tallahassee
RODNEY C. EWING, University of Michigan, Ann Arbor
JAMES H. JOHNSON, JR., Howard University, Washington, D.C.
ROGER E. KASPERSON, Stockholm Environment Institute, Stockholm, Sweden
NIKOLAY P. LAVEROV, Russian Academy of Sciences, Moscow
JANE C. S. LONG, Mackay School of Mines, University of Nevada, Reno
ALEXANDER MACLACHLAN, E.I. du Pont de Nemours & Company (retired), Wilmington, Delaware
WILLIAM A. MILLS, Oak Ridge Associated Universities (retired), Olney, Maryland
MARTIN J. STEINDLER, Argonne National Laboratory (retired), Downers Grove, Illinois
ATSUYUKI SUZUKI, University of Tokyo, Japan
JOHN J. TAYLOR, Electric Power Research Institute (retired), Palo Alto, California
VICTORIA J. TSCHINKEL, Landers and Parsons, Tallahassee, Florida

Staff

KEVIN D. CROWLEY, Director
MICAH D. LOWENTHAL, Staff Officer
BARBARA PASTINA, Staff Officer
GREGORY H. SYMMES, Senior Staff Officer
JOHN R. WILEY, Senior Staff Officer
SUSAN B. MOCKLER, Research Associate
DARLA J. THOMPSON, Senior Project Assistant/Research Assistant
TONI GREENLEAF, Administrative Associate
LATRICIA C. BAILEY, Senior Project Assistant
LAURA D. LLANOS, Senior Project Assistant
ANGELA R. TAYLOR, Senior Project Assistant
JAMES YATES, JR., Office Assistant

Acknowledgements

This study could not have been completed without the assistance of many individuals and organizations. The committee especially wishes to acknowledge and thank Inès Triay, Roger Nelson, Chuan-Fu Wu, and Mary Elisabeth "Beth" Bennington of the U.S. Department of Energy (DOE), Carlsbad Field Office. Chuan-Fu and Beth served as liaisons to the committee from the Carlsbad Field Office and ensured that all requests for documents, meetings, and other information were met with a timely response. The committee wishes to thank Kathryn Knowles and Wendell Weart from Sandia National Laboratories, who provided information and several briefings during the course of this study. Matthew Silva (Environmental Evaluation Group), Fred Ferate (U.S. Department of Transportation), Nancy Osgood (U.S. Nuclear Regulatory Commission), Robert "Bobby" Sanchez, and Mona Williams (DOE, National Transportation Program, Albuquerque Operations) were kindly available to committee members for specific clarification. The committee is grateful to all individuals who made presentations or provided information for this study.

Finally, the committee wishes to thank Barbara Pastina, Angela Taylor, and Kevin Crowley, staff of the National Research Council's Board on Radioactive Waste Management, Thomas Kiess, former staff officer and study director, and Elizabeth Ward of Garrick Consulting for great team effort to support this project.

List of Reviewers

This report has been reviewed in draft form by individuals chosen for their diverse perspectives and technical expertise, in accordance with procedures approved by the NRC's Report Review Committee. The purpose of this independent review is to provide candid and critical comments that will assist the institution in making its published report as sound as possible and to ensure that the report meets institutional standards for objectivity, evidence, and responsiveness to the study charge. The review comments and draft manuscript remain confidential to protect the integrity of the deliberative process. We wish to thank the following individuals for their review of this report:

Ray Chamberlain, Parsons Brinckerhoff, Inc.
Darleane C. Hoffman, Lawrence Berkeley National Laboratory
Leonard F. Konikow, U.S. Geological Survey
James O. Leckie, Stanford University
Harry Mandil, MPR Associates, Inc. (retired)
Michael D. Meyer, Georgia Institute of Technology
Michael O. McWilliams, Stanford University
Michael T. Ryan, Medical University of South Carolina
John J. Taylor, Electric Power Research Institute (retired)
Chris G. Whipple, Environ International Corporation

Although the reviewers listed above have provided many constructive comments and suggestions, they were not asked to endorse the conclusions or recommendations nor did they see the final draft of the report before its release. The review of this report was overseen by Frank L. Parker, Vanderbilt University. Appointed by the National Research Council, he was responsible for making certain that an independent examination of this report was carried out in accordance with institutional procedures and that all review comments were carefully considered. Responsibility for the final content of this report rests entirely with the authoring committee and the institution.

Preface

This study was sponsored by the U.S. Department of Energy (DOE) Carlsbad Field Office, formerly known as Carlsbad Area Office (CAO). To accomplish this project, the National Research Council (NRC) empanelled a 15-member committee on the Waste Isolation Pilot Plant (WIPP). Committee members were chosen for their expertise in relevant technical disciplines such as nuclear engineering, health physics, chemical and environmental engineering, civil and transportation engineering, performance assessment, analytical chemistry, materials science and engineering, plutonium geochemistry, hydrogeology, rock and fracture mechanics, petroleum engineering, and mining engineering. The committee is operated under the auspices of the Board on Radioactive Waste Management of the NRC.

The first committee on the Waste Isolation Pilot Plant was formed in 1978, at the request of the DOE, to provide scientific and technical evaluations of the investigations at the WIPP. That committee functioned as a standing committee until late 1996 at which time it published its final report (NRC, 1996a), *The Waste Isolation Pilot Plant, A Potential Solution for the Disposal of Transuranic Waste*.[1] This was the last report of the committee prior to certification of the site. The committee concluded that "human exposure to radionuclide releases from transuranic waste disposed in the WIPP is likely to be low compared to U.S. and international standards." The report went on to say, "The only known possibilities of serious release of radionuclides appear to be from poor seals or some form of future human activity that results in intrusion into the repository." The report recommended that "speculative scenarios of human intrusion should not be used as the sole or primary basis on which to judge the acceptability of the WIPP (and, by extension, any geological repository)."

Following the publication of the 1996 report, this WIPP committee was created to carry out the statement of task reported in Sidebar P.1. The committee has produced two reports to cover the statement of task, an interim report published in April 2000 and this final report. The complete interim report has been reproduced as Appendix A1.

---

[1] Transuranic (TRU) waste is waste contaminated with alpha-emitting radionuclides of atomic number greater than 92 and half-lives greater than 20 years in concentrations greater than 100 nanocuries per gram. For more details see Sidebar 1.2 and the glossary.

> **Sidebar P.1 Statement of Task**
>
> The purpose of this study is to identify the limiting technical components of the WIPP program, with a twofold goal of (i) improving the understanding of long-term performance of the repository and (ii) identifying technical options for improvements to the National Transuranic (TRU) Program (i.e., the engineering system that defines TRU waste handling operations that are needed for these wastes to go from their current storage locations to the final repository destination) without compromising safety.
>
> To accomplish this goal, the study will address two major issues:
>
> 1. The first is to identify research activities that would enhance the assessment of long-term repository performance. This study would examine the performance assessment models used to calculate hypothetical long-term releases of radioactivity, and would suggest future scientific and technical work that could reduce uncertainties.
>
> 2. The second is to identify areas for improvement in the TRU waste management system that may increase system throughput, efficiency, cost effectiveness, or safety to workers and the public. This study will examine, among other inputs, the current plans for TRU waste handling, characterization, treatment, packaging, and transportation.

In October 2000, the DOE provided a comprehensive response to the recommendations made in the interim report. The response is reported in Appendix A2. The committee is very encouraged by the quality of the responses and the actions the DOE is taking. Although the responses will not have a full impact on this final publication because of the report's tight schedule, the committee has been able to acknowledge a number of them in this report.

As is the normal practice of the National Academies, committee members do not represent the views of their institutions but form an independent body to author the report using the information gathered together with their collective knowledge and experience. The report reflects a consensus of the committee and has been reviewed in accordance with NRC procedures.

B. John Garrick, *Chair*
Committee on the Waste Isolation Pilot Plant
April 2001

Contents

EXECUTIVE SUMMARY ... 1
   Site Performance, 2
   Site Characterization, 3
   The National TRU Program, 4
      Waste Characterization and Packaging, 4
      Waste Transportation, 5

1 INTRODUCTION ... 7
   Site Performance and Characterization, 15
   The National Transuranic Program, 16

2 REPOSITORY PERFORMANCE CONFIRMATION ... 20
   Regulatory Requirements for Monitoring, 21
   Site Performance Issues, 22
      Brine Migration and Moisture Access to the Repository, 22
      Gas Generation in the Repository, 23
      Magnesium Oxide Backfill, 25
      Salt Healing and Disturbed Rock Zone Integrity, 26
   Site Characterization Issues, 27
      Geohydrological Characterization of the Rustler Formation, 27
      Oil, Gas, and Mineral Production, 28
      Baseline Radiogenic Analysis of Subsurface Fluids, 31

3 NATIONAL TRANSURANIC WASTE MANAGEMENT PROGRAM ... 33
   Waste Characterization and Packaging, 33
      Waste Characterization and Packaging Requirements, 33
      Total Inventory of Organic Material in the Repository, 34
   Waste Transportation, 34
      DOE's Communication and Notification Program, 35
      DOE's Emergency Response Program, 37
      Rail as a Transportation Option for Certain TRU Waste, 38
      Gas Generation Safety Analysis for TRUPACT-II Containers, 40

4 SUMMARY   42
   Overarching Finding, 42
   Overarching Recommendation, 42

REFERENCES   44

APPENDIXES
   A1. Interim Report   51
   A2. DOE's Response to the Interim Report   109
   B. Human Intrusion Scenarios   123
   C. Biographical Sketches of Committee Members   128
   D. Glossary   132
   E. Acronyms and Symbols   137
   F. Other Relevant Documents   139

Executive Summary

The Waste Isolation Pilot Plant (WIPP) is a deep underground mined facility for the disposal of transuranic waste resulting from the nation's defense program. Transuranic waste is defined as waste contaminated with transuranic radionuclides with half-life greater than 20 years and activity greater than 100 nanocuries per gram. The waste mainly consists of contaminated protective clothing, rags, old tools and equipment, pieces of dismantled buildings, chemical residues, and scrap materials. The total activity of the waste expected to be disposed at the WIPP is estimated to be approximately 7 million curies, including 12,900 kilograms of plutonium distributed throughout the waste in very dilute form. The WIPP is located near the community of Carlsbad, in southeastern New Mexico. The geological setting is a 600-meter thick, 250 million-year-old saltbed, the Salado Formation, lying 660 meters below the surface.

The National Research Council (NRC) has been providing the U.S. Department of Energy (DOE) scientific and technical evaluations of the WIPP since 1978. This is the first full NRC report issued following the certification of the facility by the U.S. Environmental Protection Agency (EPA) on May 18, 1998. An interim report was issued by the committee in April 2000 and is reproduced in this report as Appendix A1. The main findings and recommendations from the interim report have been incorporated into the body of this report.

The committee's task is twofold: (1) to identify technical issues that can be addressed to enhance confidence in the safe and long-term performance of the repository and (2) to identify opportunities for improving the National Transuranic (TRU) Program for waste management, especially with regard to the safety of workers and the public. The complete statement of task is reported in Sidebar P.1 of the Preface.

The **overarching finding and recommendation** of this report is that the activity that would best enhance confidence in the safe and long-term performance of the repository is to monitor critical performance parameters during the long pre-closure phase of repository operations (35 to possibly 100 years). Indeed, in the first 50 to 100 years the rates of important processes such as salt creep, brine inflow (if any), and microbial activity are predicted to be the highest and will be less significant later. The committee recommends that the results of the on-site monitoring program be used to improve the performance assessment for recertification purposes. These results will determine whether the need for a new performance assessment is warranted. For the National TRU Program, the committee finds that the DOE is implementing many of the recommendations of its interim report. It is important that the DOE continue its efforts to improve the packaging, characterization, and transportation of the transuranic waste.

The committee's **specific findings and recommendations** have been grouped into three categories: (1) site performance, (2) site characterization, and (3) the National TRU Program.

SITE PERFORMANCE

Every five years, the WIPP must obtain recertification from the EPA by showing that the repository is performing as predicted. Site performance refers to activities, phenomena, or events that occur as a result of repository construction and waste emplacement in the time frame between placement of the waste and final sealing[1] of the repository shaft. Site performance has been evaluated by the DOE in its Compliance Certification Application (CCA) (DOE, 1996). The CCA relies on a model, called a "performance assessment," that calculates the probability and consequence of several scenarios by which radionuclides could be released into the environment. The performance assessment also identifies the major uncertainties and their impact on the overall performance of the system. To reduce some of the uncertainties in the performance assessment and to add confidence in the containment performance of the repository, *the committee recommends taking advantage of the long (35 to possibly 100 years) pre-closure operating period to monitor selected performance indicators,* including those listed below:

1. Brine migration is a key issue because it provides the most realistic mechanism for mobilizing and transporting radionuclides from the waste. The mixing of brine and waste could also result in the generation of gas in the repository. **The committee recommends pre-closure monitoring to gain information on brine migration and moisture access to the repository. Observation should continue at least until the repository shafts are sealed and longer if possible. The committee recommends that the results of the on-site monitoring program be used to improve the performance assessment for recertification purposes.**

2. Gas pressure generation is an important issue in the assessment of human intrusion scenarios. In the committee's opinion, there are uncertainties in some of the assumptions about gas generation used in the performance assessment of the CCA. **The committee recommends pre-closure monitoring of gas generation rates, as well as of the volume of hydrogen, carbon dioxide, and methane produced. Such monitoring could enhance confidence in the performance of the repository, especially if no gas generation is observed. Observation should continue at least until the repository shafts are sealed and longer if possible. The results of the gas generation monitoring program should be used to improve the performance assessment for recertification purposes.**

3. Magnesium oxide (MgO) is used as backfill in WIPP to provide some control of the chemical environment of the waste and, to a lesser extent, to fill voids in the disposal locations, thus enhancing the healing process. The chemical performance of MgO depends on gas generation and brine inflow as well as other chemical processes taking place in the repository. The committee finds that there is uncertainty about the effectiveness of MgO in controlling the chemical environment of the waste. Therefore, **the**

---

[1] The terms "sealing" and "healing" are both used in this report in relation to the repository. Repository sealing refers to the emplacement of engineered barriers preventing access or leakage to and from the repository. Repository healing indicates a natural process by which the mined salt creeps in around the waste to fill all the void spaces in the repository. See also "salt creep" and "engineered barriers" in the Glossary.

committee recommends that the net benefit of MgO used as backfill be reevaluated. The option to discontinue emplacement of MgO should be considered.

4. Deformation of rock salt and interaction of salt with TRU waste containers are of interest as a part of the pre-closure performance confirmation. The creep of salt is expected to entomb the waste drums in 100 to 150 years; thus, the radionuclide mobility values used in the performance assessment might have been overestimated. This implies less migration of radionuclides from the repository into the environment. **The committee recommends pre-closure monitoring of the status of room deformation and of the disturbed rock zone[2] (DRZ) healing. Seal performance should also be assessed. Observation should continue at least until the repository shafts are sealed and longer if possible. The results of the monitoring of room deformation and DRZ healing should be included in the PA and used for recertification purposes.**

SITE CHARACTERIZATION

The WIPP program has engaged in a comprehensive program of site characterization that, in general, has been adequate to support certification of the facility. *The committee identified four areas in which additional site characterization or monitoring is recommended.* The four site characterization programs are described below:

1. A program for the hydrologic characterization of the Culebra, the most transmissive unit in the Rustler Formation. The Culebra could provide a pathway for the release of radionuclides into the environment in the event of human intrusion. **The committee recommends a monitoring program to characterize the geohydrology of the Culebra Dolomite. Tests and measurements that should be considered include angled boreholes, natural gradient tracer tests, and additional pump or injection tests. These new data should be used to confirm, or modify, the conceptual and numerical models now proposed as reasonable simulation of the actual system.**

2. A program for the detection of deep brine reservoirs below the waste disposal horizon. To improve site characterization and increase confidence in repository performance in view of the recertification application, **the committee recommends the use of seismic survey techniques for detecting large brine reservoirs below the repository.[3] In case a brine reservoir were found beneath the WIPP and its size were larger than what is already taken into account in the PA, then the DOE should conduct an extensive review of the impact of such reservoir on the repository performance. A basis would then exist to take appropriate action to ensure the safety of the repository.**

3. A program for monitoring oil, gas, and mineral production in the area. Oil, gas, and mineral extraction activities in the vicinity of the repository could threaten its integrity. **The committee recommends the development of a database to collect information on drilling, production enhancement,**

---

[2] The disturbed rock zone is the zone around an excavation, in the host rock salt, where the stress field has been modified sufficiently to cause the formation of microfractures in the rock salt.

[3] The committee recognizes that small brine reservoirs, including brine occurring as a saturated continuum, could not be detected by seismic surveys, or other noninvasive remote sensing techniques.

mining operations, well abandonments, and unusual events (accidents and natural events) in the vicinity of the WIPP site.

4. A program for establishing the baseline for naturally occurring radioactive material (NORM) in subsurface brines and hydrocarbons in the vicinity of the site. The NRC interim report recommended that the DOE develop and implement a plan to sample oil-field brines, petroleum, and solids associated with current hydrocarbon production to identify the background concentrations of naturally occurring radioactive material in the vicinity of the WIPP site, for baselining purposes. In response to this recommendation, the DOE has started to collect data and is developing a database on NORM. **The committee recommends that the DOE continue the implementation of its plan to sample oil-field brines, petroleum, and solids associated with current and future hydrocarbon production, as necessary to assess the magnitude and variability of NORM in the vicinity of the WIPP site for baselining purposes.**

THE NATIONAL TRU PROGRAM

The National TRU Program, administered by the DOE Carlsbad Field Office, is a program to accommodate all applicable external regulations and internal requirements that are associated with the characterization, certification, packaging, and transportation of TRU waste to the WIPP facility. The committee addressed two main issues pertaining to the National TRU Program: (1) waste characterization and packaging and (2) waste transportation.

Waste Characterization and Packaging

The committee reviewed some of *the waste characterization and packaging requirements* established by the National TRU Program from a safety and cost point of view. This issue was detailed in the committee's interim report. A new issue concerning the *total inventory of organic material allowed in the repository* surfaced after the committee visited the WIPP site.

1. Waste characterization and packaging requirements. A principal finding of the interim report (Appendix A1) was that many requirements and specifications having to do with waste characterization and packaging lacked a safety or legal basis. In fact, the committee concluded that some of the requirements penalized the program by adding unnecessary costs and safety risks. Examples of self-imposed waste characterization requirements are waste repackaging to dilute the hydrogen-producing components and visual examination to verify the content of waste drums and avoid miscertifications. Therefore, the committee recommended in the interim report that the DOE should eliminate self-imposed waste characterization requirements that lack a safety or legal basis. The DOE has responded to this recommendation by initiating a review of all waste characterization and packaging requirements (Appendix A2). **The committee recommends that the DOE's efforts to review waste characterization and packaging requirements continue and that changes be implemented over the entire National TRU Program. The committee recommends that the resources required to complete these improvements be made available by the DOE.**

2. Total inventory of organic materials allowed in the repository. Buried with the waste is a considerable inventory of organic materials, such as plastic film used to stabilize the drums, plastic bags and corrugated cardboard, wooden waste boxes, plastic liners of waste drums, and pressed wood "slip sheets."

The principal concern of the committee is that the DOE does not appear to keep an accurate inventory of such organic material. **The committee recommends a risk-based analysis of the total organic material regulatory limits in WIPP. If accounting for the organic material is important to the safety of the repository, an inventory record system should be implemented as soon as possible to provide a basis for meaningful safety analysis.**

Waste Transportation

The committee has examined various aspects of the WIPP TRU waste transportation system, focusing on system safety and the cost-effectiveness of planned and ongoing activities. In its interim report (Appendix A1), the committee reviewed DOE's TRANSportation Tracking and COMmunication (TRANSCOM) system and its emergency response program. In addition to the *DOE's communication and notification program* and its *emergency response training,* two other issues have been revisited in this report: the potential use of *rail as a transportation option for certain TRU waste,* and *gas generation safety analysis for Transuranic Package Transporter, Model II (TRUPACT-II) containers.*

1. DOE's communication and notification program. The committee's interim report (Appendix A1) reviewed the transportation system for WIPP waste and particularly addressed the issue of the DOE's communication and notification system TRANSCOM and its emergency response program. The committee raised questions about the reliability and ease of use of the TRANSCOM system. Meanwhile, the DOE appears to be moving systematically toward the implementation of an efficient, comprehensive, and state-of-the-art communication and notification system, called TRANSCOM 2000. **The committee recommends that the DOE implement as soon as possible the new TRANCOM 2000 communication and notification system. Moreover, because the human factor is an important element of transportation system quality, TRANSCOM 2000 should include methods to minimize the occurrence and impact of human errors.**

2. DOE's emergency response training. Although the committee is aware of the fact that the DOE is not directly responsible for the emergency response program, DOE should nevertheless identify the resources (e.g., responders, medical facilities, recovery equipment, response teams) that might be necessary to respond to a transportation incident. **The committee recommends that the DOE facilitate the involvement of states in developing and maintaining an up-to-date, practical, and cost-effective spatial information database system to coordinate emergency responses. The DOE should also develop an ongoing assessment program for states' emergency response capabilities and allocate training resources to address deficiencies in coverage along WIPP routes.**

3. Rail as a transportation option for certain TRU waste. Among the generator sites, some have rail-loadings and tracking capabilities that could be used for railway shipping of TRU waste to WIPP. The objective of the following recommendation is to minimize the number of road shipments, and therefore the related risk, and to optimize the waste load for shipments of inner waste packages that are unsuitable for placement in TRUPACT-II overpacks. **The committee recommends that all reasonable transportation options including reduction in the number of shipments, such as rail and road transportation with better-adapted containers, should be part of the decision-making process of transporting TRU waste from generator and storage sites to the WIPP. Future transportation studies should consider railway shipments and their impact on both the safety and the cost of the program. The**

DOE should also continue to pursue the development of packaging alternatives for materials not suitable for TRUPACT-II containers.

4. Gas generation safety analysis for TRUPACT-II containers. Hydrogen gas is generated in the shipping containers by radiolytic decomposition of the organic materials in waste during transportation of TRU waste to the WIPP. The root issue is the interpretation of the U.S. Nuclear Regulatory Commission's (USNRC's) regulations on shipments involving possible flammable gases. The questions of interpretation center around the allowed volume fractions of flammable gases and the definition of the confinement barrier. Depending on interpretation, the regulations can become a severe constraint on TRU waste shipments, with no apparent benefit. In particular, the committee was unable to verify the technical basis for some of the interpretations of the regulations as they relate to the safety of the workers and the public. **The committee recommends a risk-informed analysis of WIPP specific shipment issues to identify core problems related to hydrogen generation and, perhaps, provide a basis for alternative cost-effective criteria while reducing the risk. The committee recommends the use of such risk-informed analysis in the application for revision of the USNRC certificate of compliance concerning hydrogen generation limits for transportation purposes.**

1

Introduction

The Waste Isolation Pilot Plant (WIPP) is the world's first deep underground operational geological repository for the disposal of radioactive waste. The WIPP consists of an underground mined facility located in a 250 million-year-old bedded salt formation (the Salado Formation), which lies 660 meters below the surface in a semiarid desert near the community of Carlsbad, New Mexico. The WIPP repository has been established for the disposal of transuranic (TRU) waste resulting from the nation's defense program. The advantages of the WIPP as a transuranic waste disposal are listed in Sidebar 1.1. Figures 1.1-1.3 show the location, layout, and geologic stratigraphy of the WIPP.

Transuranic waste contains alpha-emitting radionuclides that have atomic numbers greater than 92, the atomic number of uranium, the heaviest natural element. The WIPP Land Withdrawal Act (LWA) (U.S. Congress, 1992) defined TRU waste as waste contaminated with transuranic radionuclides with half-life[1] greater than 20 years and activity greater than 100 nanocuries per gram. It mainly consists of contaminated protective clothing, rags, old tools and equipment, pieces of dismantled buildings, chemical residues, and scrap material. Tables 1.1 and 1.2 provide, respectively, the inventory of major radionuclides in the WIPP and the repository inventory by waste category. More details on transuranic waste are given in Sidebar 1.2. Figure 1.4 shows pictures of typical TRU waste. Even though the backfill magnesium oxide (MgO) appears in the repository inventory, it is not considered to be waste. Water is also not part of the waste inventory. There is only a negligible amount of water in the waste, mostly water vapor and less than 1 volume percent of free liquids as allowed by the Waste Acceptance Criteria (DOE, 1999).

Packed in 55-gallon steel drums and wooden boxes, TRU waste is currently being stored at various sites across the nation. The source of the waste is the manufacture of nuclear warheads and the cleanup of the nuclear weapons sites. The risks associated with transuranic waste are related primarily to plutonium. Plutonium's long half-life (24,000 years for plutonium-239)[2] and toxicity must be considered in assess-

---

[1] The half-life is the time required for half of the atoms of a radioactive substance to disintegrate.

[2] Plutonium-239 indicates the isotope of mass number 239 of the element plutonium. The same notation is used for other radionuclides throughout this report.

Sidebar 1.1 Why the WIPP?

The rationale for isolating nuclear wastes through deep geologic disposal is based on a large body of U.S. and international research. The National Academy of Sciences observed in 1957 (NRC, 1957): "The best means of long-term disposal . . . is deep geological emplacement. . . ." The Academy reaffirmed and expanded on this view in NRC (1984) and in NRC (1996a). The WIPP repository is carved out of a bedded salt formation, with the following features that make it ideal for transuranic waste disposal:

Dry environment. Large salt beds are found only in geologic regions that lack significant flows of groundwater. This deep, relatively dry underground environment greatly reduces the possibility that wastes could be carried out of a repository by natural processes. The saltbed at the WIPP site has been stable for 225 million years. It can be expected, with high confidence, to remain that way for many thousands of years into the future.

Waste immobilization. Salt tends to "heal" itself after being mined because it gradually creeps under the pressure from overlying earth and fills any openings. After several hundred years, the salt at the WIPP is expected to close in on the waste and lock it deep below the surface.

Since the mid-1970s, the Department of Energy (DOE) and its scientific adviser, Sandia National Laboratories, have studied the WIPP site to make sure it is a safe place to isolate transuranic waste. The WIPP addresses the following two key national needs:

Reducing risk. As long as transuranic waste remains at storage sites, there will be some level of risk to populations near these sites. Also, workers who must maintain current sites and monitor wastes are frequently exposed to low levels of radiation.

Providing disposal. The WIPP is a first-of-its-kind deep geologic disposal facility and will provide a model for radioactive waste disposal. In addition to the existing inventory of stored transuranic waste, estimated at about 2.32 million cubic feet, the WIPP will be the disposal site for more than 3.7 million cubic feet of transuranic waste expected to be generated during the next 35 years as DOE sites are closed. Under current law, the DOE is allowed to store 6.2 million cubic feet of transuranic waste at the WIPP. SOURCE: Citizens' Guide to the Compliance Certification Application (DOE, 1996b).

SOURCE: Citizens' Guide to the Compliance Certification Application (DOE, 1996b).

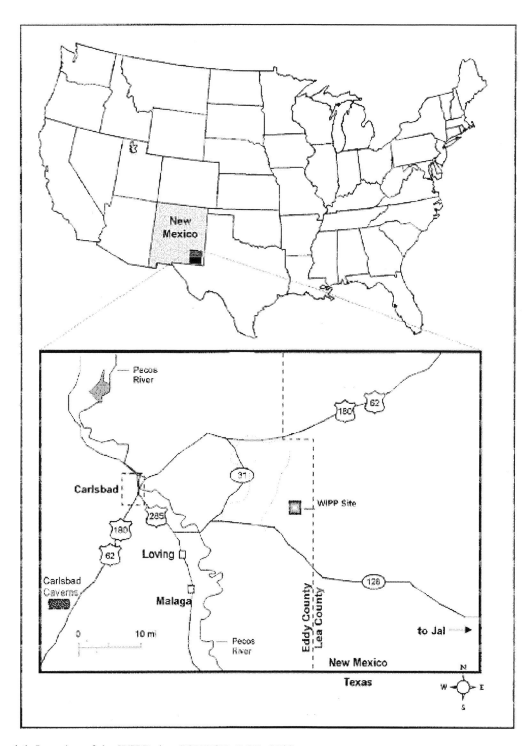

Figure 1.1 Location of the WIPP site. SOURCE: DOE, 2000g.

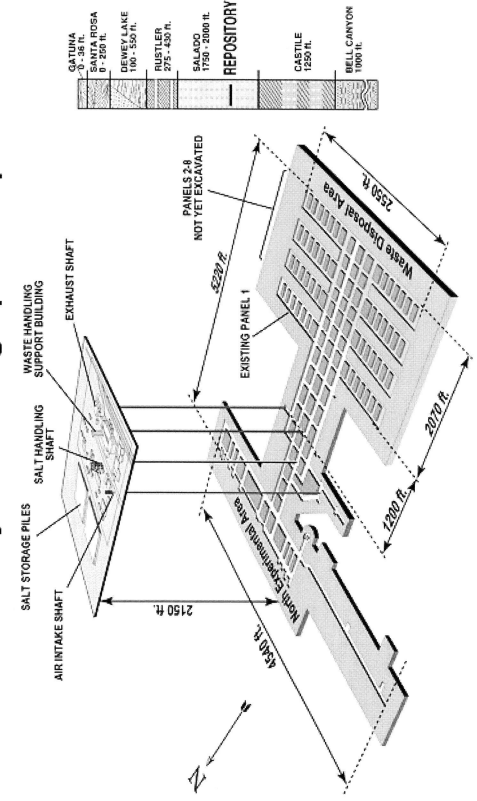

Figure 1.2 The WIPP facility and stratigraphic sequence. Panel 1 is currently in use. The mining of Panel 2 was completed on October 13, 2000. SOURCE: DOE, 2000h.

INTRODUCTION

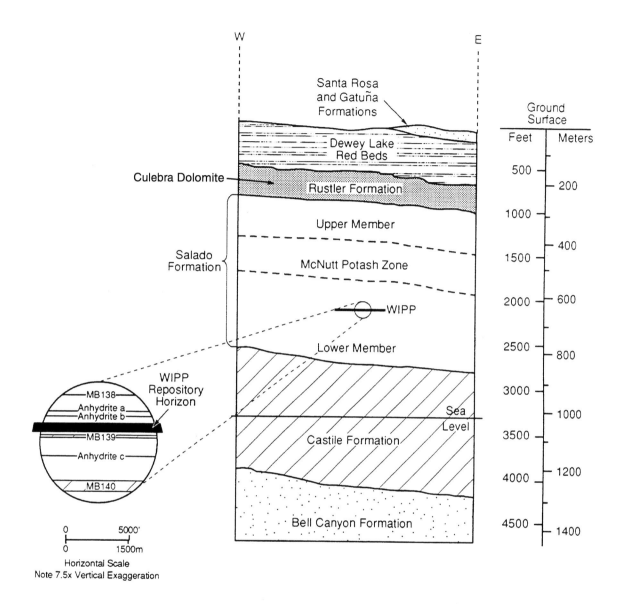

Figure 1.3 WIPP stratigraphy and depths of four key formations (Castile Formation, Salado Formation, Rustler Formation, and Dewey Lake Red Beds) including the position of the WIPP repository within the Salado. The Culebra Dolomite is one of the members of the Rustler Formation. It is approximately 7-8 meters thick at the WIPP site. Because it is a relatively transmissive unit, the Culebra is important to the groundwater flow model for the WIPP site. Inset shows finer-scale stratigraphy around the repository horizon, with marker beds and other thin beds. Adapted from Jensen et al. (1993).

Table 1.1 Inventory of the Most Abundant Radionuclides Expected in the Repository.[a]

Radionuclide	Contact Handled (CH)-Transuranic (TRU) Waste (Ci/m^3)	Remote Handled (RH)-Transuranic (TRU) Waste (Ci/m^3)
Am-241	2.62	0.842
Ba-137m	4.53×10^{-2}	28.9
Cm-244	0.187	4.45×10^{-2}
Co-60	3.83×10^{-4}	1.47
Cs-137	4.78×10^{-2}	30.5
Pu-238	15.5	0.205
Pu-239	4.66	1.45
Pu-240	1.25	0.715
Pu-241	13.7	20.0
Sr-90	4.07×10^{-2}	29.5
Y-90	4.07×10^{-2}	29.5

[a]The expected volumes of CH waste and RH waste are, respectively, 160,000 and 7,079 cubic meters.
SOURCE: DOE, 1996.

Table 1.2 Repository Inventory by Waste Category

Waste Category	Inventory (wt%)
Iron-based metal, alloys	14
Steel container material	12
Aluminum-based metal, alloys	1
Other metal, alloys	6
Other inorganic materials	3
Vitrified	5
Cellulosics	4
Rubber	1
Plastics	3
Plastic container or liner material	2
Solidified inorganic material (including cement)	4
Solidified organic material (not including cement)	0
Solidification cement	4
Soils	4
MgO backfill	37

SOURCE: Knowles et al., 2000.

ing not only the long-term risk of the WIPP, but also the potential radiation exposure of workers who handle, repackage, and transport the waste.

The WIPP has been under study since the mid-1970s, began construction in January 1981, was certified by the U.S. Environmental Protection Agency (EPA) in May 1998, and received its first transuranic waste shipment from the Los Alamos National Laboratory in March 1999. The first out-of-state shipment was received in June 1999 from the Rocky Flats Environmental Technology Site, and in September 2000, the first mixed-waste shipment was received from the Idaho National Engineering and Environmental Laboratory (INEEL). Figure 1.5 shows the main waste generators and the transportation routes to the WIPP.

> **Sidebar 1.2 What Is TRU Waste And How Is It Classified?**
>
> Transuranic waste is waste that contains alpha particle-emitting radionuclides with atomic numbers greater than that of uranium (92), half-lives greater than 20 years, and concentrations greater than 100 nanocuries per gram of waste. TRU waste is classified according to the radiation dose rate at package surface. As defined in the LWA, **contact-handled** (CH) TRU waste has a radiation dose rate at package surface not greater than 200 millirem per hour; this waste can safely be handled directly by personnel. **Remote-handled** (RH) TRU waste has a radiation dose rate at package surface of 200 millirem per hour or greater, but not more than 1,000 rem per hour (U.S. Congress, 1992); this waste must be handled remotely (i.e., with machinery designed to shield the handler from radiation). Alpha radiation is the primary factor in the radiation health hazard associated with TRU waste. Alpha radiation is not energetic enough to penetrate human skin but poses a health hazard if it is taken into the body (e.g., inhaled or ingested). In addition to alpha radiation, TRU waste also emits gamma and/or beta radiation, which can penetrate the human body and requires shielding during transport and handling. RH TRU waste has gamma and/or beta radiation-emitting radionuclides in greater quantities than exist in CH TRU waste (DOE, 2000a).
>
> TRU waste is further classified as TRU waste or **mixed** TRU waste. Mixed TRU waste contains both radioactive materials regulated under the Atomic Energy Act and hazardous chemical compounds regulated under the Resource Conservation and Recovery Act.
>
> The total activity of the waste expected to be disposed at the WIPP is estimated to be approximately 7 million curies (of which 6 million is from CH waste), including 12,900 kilograms of plutonium distributed throughout the waste in very dilute form. According to the Compliance Certification Application (CCA), the volume of CH waste expected in WIPP is 160,000 cubic meters and that of RH waste is 7,079 cubic meters (DOE, 1996).

The WIPP is designed to dispose of approximately 175,000 cubic meters of transuranic waste. Total activity of the waste is estimated to be approximately 7 million curies. The largest fraction of this activity comes from approximately 12,900 kilograms of plutonium distributed throughout the waste in very dilute form. TRU waste is classified as contact-handled (CH) and remote-handled (RH) waste, according to the radioactivity at the container surface [3] (see Sidebar 1.2). According to the National TRU Waste Management Plan, the disposal of RH waste will not begin before early 2002 (DOE, 2000a). Since most of the radioactivity is coming from the plutonium in CH waste (approximately 85 percent of the total curies inventory, see Table 1.1), the disposal of RH waste should not represent a significant added risk to the repository. A further issue concerning RH waste will be discussed in relation with the emplacement of backfill in Chapter 2.

This report presents the results of a National Research Council (NRC) study of operational, technical, and programmatic issues associated with the long-term performance of the WIPP. Previous studies

---

[3] This type of classification is intended for the protection of workers handling radioactive waste. Public health protection standards have also been taken into account in the design and operation of the WIPP.

14 IMPROVING OPERATIONS AND LONG-TERM SAFETY OF THE WASTE ISOLATION PILOT PLANT

Figure 1.4 Radiography of a transuranic waste drum. SOURCE: DOE, 2000i.

by the NRC's committee on the WIPP covered ongoing activities in preparation for the opening of the facility. This study is the first to address the WIPP as an operational repository.

The seeds for this report were planted during the preparation of the 1996 report by the previous WIPP committee (NRC, 1996a). That committee observed that the long operating period of the WIPP (at least 35 years and possibly much longer) provides an opportunity to conduct studies and investigations that would decrease some of the uncertainties about the long-term safety performance of the repository.

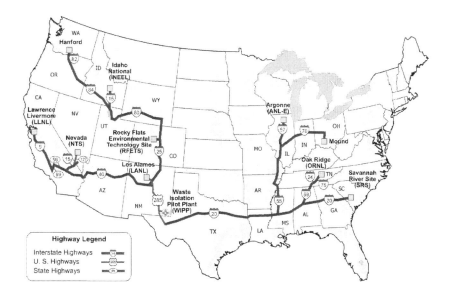

Figure 1.5 Defense transuranic waste generating and storage sites and primary transportation routes. SOURCE: DOE, 2000j.

Thus, this committee has focused on identifying studies and investigations "that would enhance the assessment of long-term repository performance," as noted in the statement of task in the Preface to this report. The second part of this committee's statement of task addresses potential improvements to the National Transuranic Waste Management Plan, also known as the National TRU Program. This program coordinates the management and disposal activities of TRU waste between the WIPP and the 23 generator sites. As written in the statement of task, the committee must "identify areas for improvement in the TRU waste management system that may increase system throughput, efficiency, cost-effectiveness, or safety to workers and the public." The result is the consideration of issues having to do with waste characterization, packaging requirements, waste transportation and handling, and emergency preparedness.

The two-part statement of task required very different skills and approaches: the first part is related to site performance, while the second is programmatic. The committee has chosen to structure this report into two primary sections that can be mapped directly to the two principal requirements of the statement of task. The part of the statement of task relevant to the long-term performance of the repository is addressed in the context of the repository performance confirmation program, in reference to enhancing confidence in the performance assessment models. The task relating to programmatic issues is addressed in the context of the National TRU Program.

SITE PERFORMANCE AND CHARACTERIZATION

To evaluate the long-term performance of the disposal system, the DOE uses a technique developed especially for predicting the behavior of geologic repositories over the thousands of years required for waste isolation. This technique is called "performance assessment." Performance assessment (PA) is a multidisciplinary, iterative, analytical process that begins by using available information that characterizes the waste and the disposal system (the design of the repository, the repository seals, and the natural barriers provided by the host rock and the surrounding formations). To obtain certification for the WIPP, the DOE used the PA tool to estimate the releases of radionuclides within the first 10,000 years, based on the probabilities of relevant features, events, and processes occurring.

The performance of the repository has been assessed for two main scenarios: the undisturbed repository scenario and the human intrusion scenario. If the repository is left undisturbed, the only release pathway for radionuclide release into the environment is through leakage of brines containing radioactive materials into the environment. Scenarios for the disturbed case involve releases resulting from boreholes drilled inadvertently into the waste. According to the Land Withdrawal Act (U.S. Congress, 1992), the DOE must exercise active institutional controls[4] on a perimeter of land extending up to 5 kilometers from the boundaries of the WIPP site for 100 years after the closure of the repository. During this period, there will be no natural resource extraction activities in the site. Between 100 and 700 years after the closure of the repository, the site will be under passive institutional controls.[5] During this period, drilling activity is expected to resume and to reach its maximum after 700 years, when the land will be released to public use and the WIPP site will be no longer controlled. Uncontrolled extraction activities would increase the probability of drilling directly into the repository.

Sensitivity analyses are used by the DOE to determine which parameters of the disposal system exert the greatest effect on performance (DOE, 1996). Performance assessment calculations show that in the absence of human intrusion, brine inflow and gas generation are the most important parameters affecting the performance of the WIPP (Helton, 2000d). In the case of the disturbed scenario, the most important parameter is the borehole permeability (Helton, 2000e). Sidebar 1.3 describes the main results of the performance assessment and their implication for the long-term performance of the WIPP. For a complete review of the PA for the WIPP see Apostolakis et al. (2000). The containment requirements are set by the regulatory agency, the U.S. Environmental Protection Agency, and are listed in Sidebar 1.4. More information on regulatory compliance can be found in the previous NRC report on the WIPP (NRC, 1996a).

The EPA certified the WIPP on the basis of the performance assessment included in the Compliance Certification Application (CCA). While various mechanisms and scenarios, including their uncertainties, were considered in the performance assessment, the question now is how to enhance the degree of confidence expressed by the performance assessment results. The conceptual structure and the development of scenarios for the WIPP's PA are described respectively in reference Helton et al. (2000a) and Galson et al. (2000).

The uncertainties in the PA for the WIPP are analyzed in Helton et al. (2000b,c). The current committee on the WIPP believes that better knowledge of site performance and better site characterization are important in decreasing the uncertainties, and therefore possibly enhancing the confidence, in the performance assessment of the repository. The committee's approach to examining the PA was to focus on underlying assumptions and results of the performance assessment. Of particular interest to the committee was how the results could be impacted by uncertainties and relied upon EPA's certification for proof of the ability of the computer program to represent the model adequately. The issues and their uncertainties are discussed in Chapter 2 as site performance and site characterization issues.

THE NATIONAL TRANSURANIC PROGRAM

The National Transuranic Waste Management Plan, also known as the National TRU Program, is a plan that organizes the activities concerning storage, characterization, packaging, handling, transporta-

---

[4] Active institutional controls imply restrictions on land access or use.

[5] Passive institutional controls imply the identification of the controlled area through signs or markers; also, records are kept on the repository and land use.

Sidebar 1.3 Performance Assessment and Regulatory Acceptance

The Environmental Protection Agency's certification of the WIPP facility was based on the performance assessment submitted as a part of the U.S. Department of Energy's Compliance Certification Application. The regulatory basis for the PA for the WIPP is described in Howard et al. (2000). The PA is a computerized, mathematical model that evaluates the performance of the WIPP repository over its lifetime. The main results of this model are shown in Figure A below, and are compared there with the acceptance criterion established by the EPA shown as the line in the upper right corner. These requirements are reported in Sidebar 1.4. The horizontal axis is a measure C, of the total radioactivity released from the repository to the biosphere during its nominal 10,000-year lifetime. The vertical axis shows the "probability of release," that is, at any value of C, the probability that the actual release from the repository will exceed that value. Such a curve is called a "complementary cumulative distribution function (CCDF)." It expresses quantitatively the state of knowledge of the analysis team about how much radioactivity will be released from the repository over its lifetime. It is important to observe that the curve is well to the left of the regulatory acceptance boundary set by the EPA, meaning that the repository is in compliance with the regulation.

A variation on this form of presentation is shown in Figure B. In this figure, a family of CCDFs is traced to show the different effects of uncertainties arising from possible human intrusions into the repository (mainly by drilling into it) and the geotechnical uncertainties (e.g., physical and chemical properties of the salt). Again, the important result is that the whole family of curves is well to the left of the EPA acceptance boundary. In addition, the curves bunched close together indicate a reasonable bound on the uncertainties and add confidence that a substantial margin of safety exists.

continued

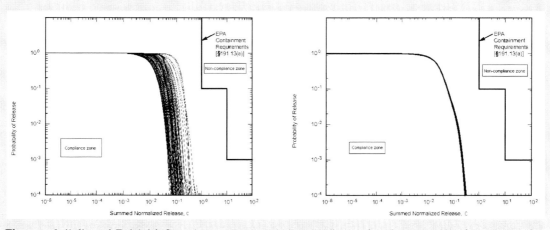

Figures A (*left*) **and B** (*right*) Complementary cumulative distribution functions resulting from the performance assessment. In A, the probability of radionuclide release from the repository is compared to the acceptance criteria. In B, a family of CCDF curves is traced to show the effect of different uncertainties. The "summed normalized release" of radionuclides C, is related to WIPP containment requirements in Sidebar 1.4. The term "normalized" release means that the release C_j is divided by the release limit L_j. The use of the term "summed" indicates the sum of all C_j/L_j over all the radionuclides with half-life greater than 20 years. The summed normalized release represents the total radioactivity released from the repository to the biosphere during its nominal 10,000-year lifetime. More details on CCDFs can be found in NRC (1996a). SOURCE: DOE, 2000k.

Sidebar 1.3 Continued

The committee recognizes that computing the performance of an underground repository over many millennia into the future cannot be done today with the accuracy with which, for example, the performance of an airplane wing can be simulated. Nevertheless, the results of this performance assessment are considered adequate by experts and regulators to support the decision to move waste from its surface storage to the WIPP (EPA, 1998).

Sidebar 1.4 Containment Requirements

Title 40 CFR 191.13 requires that "disposal systems for . . . transuranic radioactive wastes shall be designed to provide a reasonable expectation, based on performance assessments, that the cumulative releases of radionuclides to the accessible environment for 10,000 years after disposal from all significant processes and events that may affect the disposal system shall:

1. Have a likelihood of less than one chance in 10 of exceeding the quantities calculated according to Table 1 . . .; and
2. Have a likelihood of less than one chance in 1,000 of exceeding ten times the quantities calculated according to Table 1. . .".

To explain how these requirements are applied to the WIPP, let L_j be the limit shown in the above table for radionuclide j. Suppose for the moment that WIPP had only one radionuclide, j, and let C_j be the total release of that radionuclide to the environment, measured in curies per 1000 metric tons of heavy metal (MTHM), during its 10,000 year lifetime. Then the first requirement of 40 CFR 191.13 means that the probability of C_j being greater than L_j should be less than 0.1.

That is: $p(C_j/L_j >1)$ should be < 0.1

The second requirement then indicates that

$p(C_j/L_j >10)$ should be < 0.001.

The actual inventory of radionuclides C, is defined as:

$$C = \sum_{j}^{N_j} \frac{C_j}{L_j}$$

with N_j being the total number of radionuclides with a half-life greater than 20 years. The requirements then become:

$p(C>1)$ should be < 0.1
$p(C>10)$ should be < 0.001

Table 1. Release Limits per 1,000,000 Curies of TRU Waste per 10,000 Years[a]

Radionuclide	Release Limit (curies per 1000 MTHM)
Americium-241 or 243	100
Carbon-14	100
Cesium-137 or 137	1,000
Iodine-129	100
Neptunium-237	100
Plutonium-238, 239, 240, or 242	100
Radium-226	100
Strontium-90	1,000
Technetium-99	10,000
Thorium-230 or 232	10
Tin-126	1,000
Uranium-233, 234, 235, 236, or 238	100
Any other alpha-emitting radionuclide with a half-life greater than 20 years	100
Any other radionuclide with a half-life grater than 20 years that does not emit alpha particles	1,000

[a] Containment requirements for selected isotopes as declared in Title 40 CFR 191, Appendix A (EPA, 1995). The release limits specified here scale with the quantity of waste in a repository; for this reason, they are specified in terms of curies that may be released per 10,000 years per 1,000 metric tons of heavy metal (MTHM). For a repository such as WIPP, which is intended to contain transuranic wastes, EPA has established in 40 CFR 191 that 1,000 MTHM is equivalent to 1,000,000 curies of TRU wastes with greater than 20-year half-lives. Therefore, the limits specified are applicable per million curies of TRU waste.

tion, and disposal of defense-related transuranic waste to the WIPP from the 23 generator sites. The National TRU Program is administered by the DOE's Carlsbad Field Office. The goals of the National TRU Program are the following:

- achieving regulatory compliance among all the sites,
- reducing risk while maximizing rate of TRU waste disposal,
- reducing mortgage costs by closing the generators' sites as soon as possible, and
- using the WIPP effectively by coordinating the shipments with the repository's waste-handling and disposal capabilities.

The issues considered in this report relate primarily to waste characterization and packaging and waste transportation. Because of their importance in the near term for achieving the beginning of operation at the WIPP, the committee focused on these issues in its interim report, reported in Appendix A1. In Chapter 3 of this final report, the committee re-visits the issues related to characterization, packaging, and transportation of the wastes, including communication systems and emergency preparedness. The issue of hydrogen gas generation, as it applies to both waste characterization and transportation, is also discussed in Chapter 3.

2

Repository Perfomance Confirmation

The performance of geological repositories is evaluated on the basis of their ability to comply with a series of regulatory performance criteria, defined in Title 40 of the Code of Federal Regulations Part 191 (40 CFR 191; EPA, 1985). In the case of the WIPP, the time of compliance with the containment requirements formulated by the U.S. Environmental Protection Agency (EPA) is 10,000 years (see Sidebar 1.4). Of course, the issue does not end at 10,000 years; for example, the risk-controlling radionuclide in the WIPP is plutonium-239, which has a half-life of 24,000 years. The intent of a geological repository is to contain the waste for the indefinite future (e.g., > 10,000 years).

The WIPP was certified by the EPA through a comprehensive process based primarily on a detailed performance assessment (DOE, 1996; see also Sidebar 1.2). Of course, acceptance of the performance assessment (PA) is conditional on several factors that are designed to offset the many uncertainties involved. One of the EPA's requirements is that the DOE must implement a monitoring program designed to provide confidence in the assessed performance of the repository. Furthermore, every five years, the DOE must apply to the EPA for recertification of the WIPP. The recertification application must show evidence that the repository is performing as assessed.

A monitoring program that emphasizes factors contributing mostly to performance uncertainties could provide important evidence of the ability of the repository to perform its intended function. Therefore, the committee has chosen as the theme of this review to be "performance confirmation through monitoring." The strategy of the committee is to focus on safety and monitoring activities that would best enhance confidence in the long-term performance and reduce uncertainties in the performance assessment of the WIPP.

The recommendation to implement an in-situ monitoring program was endorsed by a previous NRC committee on the WIPP in a letter report to the Hon. L. P. Duffy (NRC, 1992). Quoting from that report, "The panel emphasizes that it supports the notion of underground testing with TRU wastes, provided that the underground location does not prevent important tests from being carried out (e.g., the measurement of brine compositions in contact with real waste or progression of gas generation experiments without purging), and that the tests can be continued for sufficient time to provide useful information."

The long operational phase of the WIPP repository (at least 35 years and possibly as long as 100 years) provides an unusual, and perhaps unprecedented, opportunity to implement a monitoring program and reexamine the performance assessment with information based on direct observations of the total system prior to closure of the repository. Although a time frame of 35-100 years is short compared to the 10,000-year period of compliance, the committee believes that it is long enough to develop and implement a monitoring program to observe the development of repository responses. Indeed, the rates of important processes such as salt creep, brine inflow, and microbial activity are predicted to be the highest during the first 50 to 100 years (Knowles and Economy, 2000; NRC, 1996a). If these responses confirm assumptions in the performance assessment, this will reduce uncertainty in the projections of long-term performance of the repository and could improve public confidence in the repository performance.

The ongoing DOE's monitoring program required by the EPA as part of the certification, is described in the next section. The committee strongly supports such a program but believes it could be more focused and risk-informed. The difference in focus between what is planned and what the committee suggests is also discussed.

REGULATORY REQUIREMENTS FOR MONITORING

A monitoring plan for the WIPP was included in 40 CFR 194 under the requirements for the certification of the repository by the EPA. The purpose of the monitoring plan is to confirm that the repository is performing as expected according to the model in the Certificate of Compliance Application. The DOE proposed a monitoring plan, which was accepted by the EPA in 1998 in the certification decision, to address the requirements of the regulations in 40 CFR 194.

The DOE described its monitoring program in the CCA and indicated that it would span 150 years (50 years pre-closure and 100 years post-closure). The DOE program evolved from screening 91 potentially significant parameters down to 10. The 10 parameters were divided among physical measurements in the Salado Formation, hydrological properties in the non-Salado settings, and activity levels of the waste. The four parameters to be measured in the Salado Formation relate to creep closure and stresses, extent of deformation, initiation of brittle deformation, and displacement of deformation features. The program calls for pre-closure monitoring only for the Salado parameters and pre- and post-closure for the non-Salado parameters. Waste activity is to be monitored only during pre-closure. In the DOE program, pre-closure monitoring in waste storage rooms ends with the sealing of individual panels of rooms; hence, pre-closure monitoring of emplaced waste is very limited. The parameters that the DOE is currently monitoring to comply with 40 CFR 194 are shown in Table 2.1.

The committee's proposed performance confirmation monitoring plan is very similar to the DOE's monitoring program. The significant difference between the DOE monitoring program and the committee proposal is that the committee's recommended plan includes monitoring rooms and panels after sealing of the panels and extends until closure of the repository and sealing of the shafts. The committee has put greater emphasis on such issues as brine inflow, gas generation, salt rock deformation following sealing of the panels, auxiliary material inventory in the repository, and radiogenic measurements.

The committee identified important issues relative to the long-term safe performance of the WIPP repository on the basis of the DOE's performance assessment (DOE, 1996), past committee reports, and numerous briefings on the WIPP. The criteria for identifying issues were related principally to the sources of uncertainty in the performance assessment and to the safety of workers and the public. Several of the issues are interrelated but are treated separately to emphasize important points. The following paragraphs describe in detail the issues of consideration in the performance confirmation monitoring program proposed by the committee. The issues have been grouped as *site performance* issues and *site characterization* issues.

Table 2.1 Parameters Currently Monitored by the DOE to Comply with 40 CFR Part 194.42[a]

Parameter Monitored in the WIPP	Pre-closure Monitoring?	Post-closure Monitoring?
Salado physical parameters		
Creep closure and stresses	YES	NO
Extent of deformation	YES	NO
Initiation of brittle deformation	YES	NO
Displacement of deformation features	YES	NO
Non-Salado hydrological properties		
Culebra groundwater composition	YES	YES
Probability of encountering a Castile brine reservoir	YES	YES
Drilling rate	YES	YES
Culebra change in groundwater flow	YES	YES
Subsidence measurements	YES	YES
Waste related parameters		
Waste activity	YES	NO

[a] EPA (1996).

SITE PERFORMANCE ISSUES

The key site issues that should be monitored during the pre-closure period to confirm the performance of the WIPP repository are described below.

Brine Migration and Moisture Access to the Repository

The presence or absence of brine in the WIPP rooms is a key issue in the performance of the repository. Without brine there will be no radionuclide mobilization and transport or any gas generation from corrosion of the steel drums. The brine sources for the undisturbed repository are seepage from brine-filled void spaces in the undisturbed geological setting, the humidity of the repository air, and water used during mining operations. In the long term, after repository closure, additional sources of brine could include accidental fluid injections by inadvertent human intrusions (see section "Oil, Gas, and Mineral Production" and Appendix B). Brine volumes from enhanced recovery fluid injection operations have the potential to be a source of much greater brine inflow than that expected from any other water sources in the undisturbed geological setting. A concern is the possible failure of a well casing or cement outside the casing during an injection operation and fluid leaking into overlying formations and flowing laterally along one of the several anhydrite layers in the Salado Formation.

A previous NRC committee analyzed the brine accumulation issue and concluded that "the formation of an abundant mobile fluid in a repository at the WIPP site . . . is very improbable." Nevertheless, the same committee recommended a "well conceived experimental program at WIPP to reduce remaining uncertainties" (NRC, 1988). The present committee is also in favor of a monitoring program to complement DOE's current program. It would be informative to monitor the brine flow rate into the first panel, or panels, of the WIPP facility that are filled and sealed. Monitoring from inside the face of the seal should be possible for decades after the panel is sealed and would contribute to enhancing confidence in the performance of the repository. Monitoring the humidity and the accumulation of standing brine would indicate the ingress of brine, although salt mines are notoriously dry and probably no standing water will develop. The monitoring of brine inflow and of the humidity in the WIPP should continue at least until the shafts are sealed and longer if feasible.

According to Knowles and Economy (2000), the brine inflow rate will be maximum within the first 50 to 100 years from the mining of the repository and will stabilize progressively after 200 years. The rate of brine inflow depends on the porosity of the medium. The mining of the salt in the repository creates alterations of the stress field of the surrounding rock and forms micro fractures in the salt around the excavation (disturbed rock zone, or DRZ). Compared to the intact salt, the DRZ will have an increased porosity because of all the micro fractures. Over time, the porosity of the DRZ decreases as salt creep continues, thereby decreasing brine inflow. Therefore, the monitoring of brine inflow is particularly important during the pre-closure phase.

At closure, the panel conduits for the instrumentation would be plugged permanently to ensure the sealing of the repository. Maintaining instrumentation at the repository horizon beyond closure of the shafts could be impractical, unless new technologies allow remote monitoring of the repository avoiding instrument conduits through the seals.

Recommendation: **The committee recommends pre-closure monitoring of the WIPP repository to gain information on brine migration and moisture access to the repository. Observation should continue at least until the repository shafts are sealed and longer if possible. The committee recommends that the results of the on-site monitoring program be used to improve the performance assessment for recertification purposes.**

Gas Generation in the Repository

Gas generation within the WIPP is one of the issues for consideration in the overall performance of the repository. There are two possible effects of gas in the repository. The first is a physical effect due to pressure buildup from any gas. Gas may generate sufficient pressure to eject repository materials during a human intrusion event. Gas pressure could also retard creep closure and brine inflow. Gas pressure in the repository is considered one of the main uncertainties in the PA concerning radionuclide release from the WIPP (Berglund et al., 2000; Helton et al., 2000d; Stoelzel et al., 2000; Vaughn et al., 2000). A performance assessment scenario that could cause violation of the EPA repository release limits involves ejection of waste material through a borehole. In this scenario, it is calculated that the gas pressure at the repository horizon has to be greater than approximately 8 megapascals[1] to result in ejection of cuttings, cavings, and spallings that might contain radionuclides from the repository (Berglund, et al., 2000; DOE, 1996). If the gas pressure approaches the lithostatic pressure, a radionuclide release along open fractures could result.

The second effect of gas generation in the repository is chemical. The main gaseous products potentially formed in the repository are carbon dioxide (CO_2), hydrogen (H_2), methane (CH_4), nitrogen (N_2 or

---

[1] The value of 8 megapascals is the pressure exerted by a column of brine-saturated drilling fluid at the depth of the repository (Stoelzel and O'Brien, 1996). This threshold in pressure was calculated in the PA on the basis of drilling technologies using mud. The public strongly criticized this assumption because it did not take into account the increasingly popular air drilling technology. However, the EPA analyzed the PA, performed supplementary calculations, and reached the conclusion that the repository would still be in compliance with release limits, even in the event of a human intrusion through air drilling (EPA, 1998).

various nitrous oxides), and hydrogen sulfide (H_2S) (Lappin and Hunter, 1989). Most of concerns come from H_2 and CO_2. Hydrogen is a flammable gas and, in the presence of certain amounts of oxygen or water vapor, could lead to an explosion. In the case of CO_2, its solubility in any brine seeping into the repository could lower the pH of the brine, which would increase actinide solubility (Novak and Moore, 1996) as shown later in this chapter.

The sources of gas generation in the repository are three: radiolysis, metal corrosion, and bacterial action. The PA shows that total gas production is negligible under humid conditions for all substrates.[2] The main uncertainty concerns CO_2 production by microbial degradation reactions; this uncertainty was acknowledged in the PA by assigning a probability of 0.5 to the occurrence of significant microbial activity (DOE, 1996; Larson, 2000). The previous NRC committee on the WIPP (NRC, 1996a) also concluded that gas generation will be minimal, even when microbial degradation of organic material is taken into account. Although this committee concurs with the previous NRC committee and with the DOE that there is "minimal" evidence of gas generation in the WIPP, uncertainties concerning gas generation are still present. The committee is concerned that experimental data were extrapolated from laboratory experiments performed under conditions that are not indicative of the actual environment in the repository.

For instance, in the case of gas generated by radiolysis of brine and organic materials, the majority of experiments were performed with high doses of radiation, which does not apply to TRU waste. Moreover, factors that significantly decrease the rate of radiolysis—matrix depletion, pressure, and inhibition by other chemical compounds—were not taken accurately into account (INEEL, 1998; Molecke, 1979b).

In the case of microbial degradation of cellulosic compounds, rates of gas generation were extrapolated from laboratory experiments performed under humid conditions (70 percent humidity), which are not representative of the intrinsic dryness of salt repositories (Francis et al., 1997). In the case of metal corrosion, gas will not result without brine inflow, an event strongly affected by uncertainties. Furthermore, the nature of the interactions between gas-phase chemicals, the influence of pressure, and of corrosion rates is still not well understood (Telander and Westerman, 1996).

The gas generation rate is expected to be maximum during the pre-closure period because it depends on the brine inflow rate for microbial degradation, corrosion, and radiolysis (NRC, 1996a). As shown in the previous section, the brine inflow rate is expected to be maximum at the beginning of the repository life. Therefore, it is important to monitor gas generation rates and volumes during the first 35 to possibly 100 years. Furthermore, continuous monitoring for gas could lead to the early detection of anomalous behavior of the repository.

Recommendation: **The committee recommends pre-closure monitoring of gas generation rates, as well as of the volume of hydrogen, carbon dioxide, and methane produced. Such monitoring could enhance confidence in the performance of the repository, especially if no gas generation is observed. Observation should continue at least until the repository shafts are sealed and longer if possible. The results of the gas generation monitoring program should be used to improve the performance assessment for recertification purposes.**

---

[2] A substrate is a generic material, whether it is metal, natural fiber, or plastic, that generates gas via various mechanisms described in this section.

Magnesium Oxide Backfill

In the framework of repository performance confirmation, an issue closely related to brine access and gas generation in the WIPP is the performance of magnesium oxide (MgO) used as backfill. The backfill is introduced in the rooms to fill voids in the disposal locations, thus enhancing the healing process and facilitating the encapsulation of waste in salt. The choice of MgO as backfill is based on its chemical properties in addition to its properties as backfill. If brine is present in the repository, MgO will mix with it to form a compact material that will encapsulate the waste (Berglund et al., 1996). The water uptake of MgO from the brine will result in a volume expansion and in the precipitation of salt from the brine that will heal all pathways for later brine penetration.[3] If brine is not present in the rooms, creep encapsulation of waste would not progress as readily.

The chemical role of MgO is to provide some control of the chemical environment of the waste by reacting with brine and scavenging the CO_2 potentially formed in the repository. In presence of CO_2, brine pH will be decreased by the formation of carbonic acid. In the acid pH range, soluble actinide carbonate complexes can then form, increasing actinide solubility (Novak and Moore, 1996). In presence of MgO scavenging all CO_2, the pH will remain in the alkaline range (9.2 - 9.9), where actinides are less soluble and less likely to be released into the environment. The rationale for this expected action of MgO relies on the following assumptions:

1. There would be significant inflow of brine into the repository's rooms.
2. Microbes would be present and react with organic waste material to form CO_2.
3. CO_2 would dissolve and acidify the brine by forming carbonic acid.
4. MgO would react with water in the brine to precipitate brucite [$Mg(OH)_2$].
5. Brucite would remove carbonic acid from the solution to form magnesite ($MgCO_3$) via intermediate products such as hydromagnesite [$4MgCO_3 \cdot Mg(OH)_2 \cdot 4H_2O$].
6. These reactions with MgO would maintain the pH of brine between 9.2 and 9.9.

As mentioned in the previous section, there are uncertainties concerning assumption 1 about the presence of a significant amount of brine in the rooms. Since assumption 2 relies on microbial generation of CO_2 under repository conditions, it is also affected by uncertainties (see previous section). Moreover, it is unclear whether the rates of reactions in assumptions 4 to 6 are sufficiently high to be effective. The committee has several concerns. How quickly does brucite form at 25°C (Krumhansl et al., 1999; Papenguth et al., 1999)? How quickly does brucite react with carbonate at various carbonate concentrations and brine compositions?[4] When will the water trapped in hydromagnesite be returned to the fluid phase?[5]

There are also uncertainties in other factors related to the chemical environment, including the amount and timing of brine inflow to form the MgO-based chemical buffer and the presence and effectiveness of microbes responsible for the CO_2. On the issue of brine inflow, there is evidence that the salt will creep in the rooms and fill all of the openings in 100 to 150 years (Callahan and DeVries, 1991;

---

[3] If the rate of brine inflow is too high, it is uncertain whether MgO can form a compact material around the waste.

[4] The rate to reach the compliance objective (26 mole percent MgO converted) decreases with decreasing CO_2 partial pressure (Krumhansl et al., 1999).

[5] The time to transform hydromagnesite to magnesite was reported to vary between 18-200 (Zhang et al., 1999) and 2.5-1,500 years (Krumhansl et al., 1999), depending on brine composition.

Knowles et al., 2000; Stone, 1997). The absence of void spaces should provide additional protection against extensive chemical reactions between brine and the waste in a short period of time.[6]

The use of MgO in the repository as chemical backfill raises the additional issue of its placement. Because MgO must be in close contact with the drums to better scavenge all CO_2 generated from the waste and because of the way the drums are stacked in the rooms, it is not possible to add MgO mechanically after the room is filled. MgO backfill is as a dry, granular, pelletized material packaged in bags of two different sizes: a smaller bag of about 25 pounds, called the "minisack," and a large bag weighing approximately 4,000 pounds, known as the "supersack." The minisacks are placed manually around and between the drums, and the supersacks are placed with a forklift on the top of each waste stack.

Based on a study by the DOE, it appears that emplacing MgO around the waste adds about 0.726 person-rem per year to the collective dose caused by waste handling. Given that the expected collective dose to waste-handling personnel is 14.6 person-rem per year, this corresponds to about 5 percent of the total dose incurred from waste operations (DOE, 2000b). The committee does not consider this additional dose to be significant. However, once remote-handled (RH) waste and possibly high-specific-activity waste in CH waste such as plutonium-238 or americium-241 are introduced into the repository, the exposures to personnel placing the MgO bags will be considerably increased.

Considering the uncertainties about the chemical performance of the MgO backfill, the committee questions the value of its use in the repository. The same concern was expressed by some of the peer review panels of the CCA (DOE, 1996, Chapter 9.3.2). This is especially true given the small but measurable additional radiation exposure to workers involved in MgO bags emplacement. The committee is not convinced of any major chemical advantages of the MgO backfill and, if its benefits to the long-term performance of the repository cannot be verified, the option to discontinue its use should be considered.

Recommendation: **The committee recommends that the net benefit of MgO used as backfill be reevaluated. The option to discontinue emplacement of MgO should be considered.**

Salt Healing and Disturbed Rock Zone Integrity

The period between placement of waste and closure of the repository provides a window of opportunity to monitor significant deformation of the salt and self-healing of the DRZ. The DRZ is the zone around an excavation in the host rock salt where the stress field has been modified sufficiently to cause the formation of microfractures in the rock salt. Substantial deformation of the salt will occur during the operation phase, which is important in assessing the self-sealing (healing) characteristics of the repository. After an initial period of rapid deformation (a few years to decades), the rooms are expected to deform, crush, and be entirely entombed by salt within 100 to 150 years (Callahan and DeVries, 1991; Knowles et al., 2000; Stone, 1997). Since the waste drums will be immobilized in a relatively short period of time (compared to 10,000 years of compliance), the radionuclide mobility values used in the performance assessment might have been overestimated. This implies less migration of radionuclides from the repository into the environment. In addition to the general deformation and local healing of rooms and panels, an important general healing must take place in the DRZ around rooms, panels, and shafts to achieve complete closure of the disposal region.

The effect of the DRZ around WIPP rooms and panels and around the shaft seal system is important in assessing the safety performance of the repository. A complete analysis of the performance of the

---

[6] A short period of time compared to the 10,000 years mentioned in the containment requirements (see Sidebar 1.4).

shaft seal system is given in Hansen and Knowles (2000). As with all the factors affecting the performance of the repository and because of the complexity of salt rock behavior, there is uncertainty in the timing and degree of self-healing of the DRZ needed to achieve the expected isolation in the mined regions. In the committee's opinion, there is also uncertainty concerning the behavior of rigid panel seals in the ductile salt surrounding them. Therefore, frequent monitoring during the pre-closure period and assessing the status of room deformation and DRZ healing are the best approaches for reducing the uncertainties associated with closure of the waste disposal area.

Recommendation: **The committee recommends pre-closure monitoring of the status of room deformation and DRZ healing. Seal performance should also be assessed. Observation should continue at least until the repository shafts are sealed and longer if possible. The results of the monitoring of room deformation and DRZ healing should be included in the PA and used for recertification purposes.**

SITE CHARACTERIZATION ISSUES

The committee finds that there are a number of site characterization actions that would decrease uncertainties in the long-term performance of the repository. Among these, site characterization issues related to human activities are particularly important because they constitute the major risk of radionuclide release, according to the performance assessment (NRC, 1996a; Rechard, 2000). Site characterization issues and activities are described in the sections below.

Geohydrological Characterization of the Rustler Formation

The WIPP disposal panels and rooms are located in the Salado Formation, approximately 660 meters from the ground surface, as shown in Figure 1.3. The Rustler Formation, overlying the Salado Formation, consists of five sequences (members) of thin-bedded strata. The Culebra Dolomite member, also called simply Culebra, is the second member from the bottom of the formation and is the most transmissive unit in the Rustler. Thus, the Culebra is important to the groundwater flow model for the WIPP site. The geologic and hydrologic setting of the WIPP have been thoroughly described in Corbet and Swift (2000). A detailed description of radionuclide transport in the Culebra can be found in Ramsey et al. (2000). The Culebra provides pathways for the release of radionuclides into the environment in all main human intrusion scenarios (see Appendix B).

These pathways can conceivably be developed when new wells are drilled through the Culebra. High-pressure fluids are used in the drilling of oil, gas, and injection wells to contain the flow from the high pressure in formations contacted during the drilling process. Formations at shallower depths, which tend to be at low pressure, are protected from the high-pressure drilling fluids by borehole casings. However, if the drilling intersects a pressurized brine reservoir before the borehole casing is placed, and if the pressure in the formation is unexpectedly higher than the pressure exerted by the drilling fluid, the high-pressure formation fluids could flow into the wellbore and cause an underground blowout into the Culebra. Drillers would use a blowout preventer to contain any immediate surface release of brine from the repository horizon. However, release to the Culebra could be synonymous with release to the accessible environment if there were high flow rates and little retardation[7] of radionuclides.

---

[7] Parameter that describes the ratio of the net apparent velocity of the concentration of a particular chemical species to the velocity of a non-reactive species.

All human intrusion assessment models in the PA require some retardation in the Culebra to meet the EPA's repository performance requirements. Similarly, the PA models require a low flow velocity in the Culebra. Unfortunately, these models are not based on sufficient hydrological characterization of the Culebra. There is uncertainty about flow directions, flow rates, retardation characteristics, and the amounts and location of groundwater recharge and discharge to and from the Culebra. This is due partly to uncertainties about the density, size, and spatial distribution of fractures and potential karstic features. These uncertainties can be reduced through a well-designed monitoring program.

The monitoring program should include angled boreholes to verify assumptions about vertical fractures or karst conduits; monitoring wells to check on conditions of recharge and discharge, water levels, and chemical properties. The program should also include a series of tracer tests to determine spatial flow rates of groundwater and local tracer tests, including the use of new logging technologies. Tracer tests should include suites of conservative tracers injected in differing wells to test the complexities of the flow system over and beyond those withdrawn by the LWA. The tests should span the entire preclosure phase of the repository (35 to 100 years). New data should be implemented continually into scenario models, and PA calculations should be revised as appropriate.

Recommendation: **The committee recommends a monitoring program to characterize the geohydrology of the Culebra Dolomite. Tests and measurements that should be considered include angled boreholes, natural gradient tracer tests, and additional pump or injection tests. These new data should be used to confirm, or modify, the conceptual and numerical models now proposed as reasonable simulation of the actual system.**

Oil, Gas, and Mineral Production

The oil, gas, and mineral reserves in the vicinity of WIPP are considerable. As shown in the interim report (see Appendix A1, Figure 2), there have been multiple drilling operations near the WIPP site and a future increase in production activities is expected. As indicated in the previous section, brine (or any fluid) inflow to the disposal region of the WIPP repository is a serious threat to the containment of radionuclides in the repository. Therefore, it is critical that pathways are not created by human intrusion, either intentionally or unintentionally. Such pathways would allow transport of radioactive materials from the repository to the surface or would bring water or brine in contact with the substances stored in the repository.

No human intrusion should occur during the first 100 years of the repository's life because of the active institutional controls. However, drilling activity will increase progressively during the period of passive institutional controls (100 to 700 years) and will not be controlled beyond that period. Uncontrolled extraction activities would increase the probability of drilling directly into the repository. Extraction activities can be divided into *drilling activities* and *mining activities*.

Drilling Activities

Two scenarios related to drilling activities are of particular interest to the WIPP site: the Hartman scenario and the intersection of a pressurized brine reservoir.

1. The Hartman scenario. In 1993, while drilling in the Rhodes Yates oil field located approximately 45 miles from the WIPP site, Mr. Hartman experienced a well blowout followed by an uncontrollable flow of brine to the surface (Silva, 1996). This event has come to be known as the Hartman scenario (see additional details in Appendix B, Box B.1). The reason for the blowout has not been fully determined, although there is evidence that it may have been caused by a high-pressure, water-flooding operation approximately 1 mile from the well that blew out. In oil-producing regions such as southeastern New Mexico, it is common to inject high-pressure fluids into the deep rock formations of the subsurface.

The purpose of these fluid injections is to stimulate secondary recovery of oil in partly depleted oil reservoirs (e.g., by water flooding) or to dispose of large volumes of brine produced simultaneously with oil. If there is a failure in the well casing or in the grout or cement outside the casing, fluid can leak into overlying formations and flow laterally along one of the many anhydrite layers in the Salado (NRC, 1996a). Mr. Hartman might have drilled into a hydraulic fracture possibly induced by such water-flooding operation, causing the well to blow out. Bredehoeft and Gerstle (Bredehoeft and Gerstle, 1997; Gerstle and Bredehoeft, 1997) studied the implication of the Hartman scenario for the safety of the WIPP. They argued that if there were an oilfield water-flooding operation in the vicinity of the WIPP, a large amount of brine could flow from a leaky injection well and induce a hydraulic fracture in the anhydrite (or marker bed) directly above or below the WIPP repository (see Appendix B, Box B.2). If, at some later time, another well were drilled through the repository and into this brine-filled fracture, the high-pressure brine in the fracture could flow through the borehole and flood the repository causing a release of radioactive materials. Bredehoeft's analysis was disputed by researchers at Sandia National Laboratories (SNL; Swift et al., 1997; Vaughn et al., 1998). The discussion focused on the size of this potential hydraulically induced fracture and on whether this fracture could reach the anhydride beds directly below or above the repository site.

The committee's opinion is that there are considerable uncertainties concerning both the mechanism of the Hartman scenario and its likelihood to develop at the WIPP site. For instance, if the hypothesis of a hydraulically induced fracture were valid, and the fracture would indeed extend directly below or above the repository, a surge of brine would be expected only when the drillbit penetrates the brine-filled fracture. The volume of brine inflow would not be large enough to damage the repository because hydraulic fractures have small opening widths and high internal flow resistances. Furthermore, a leaky well could not provide sufficient energy and fluid volume to cause a brine inflow into the repository for an extended period of time; also, the energy stored in the room and in the fracture would not be enough to push the waste to the surface.

In addition, the repository is partitioned into isolated rooms, which will be closed progressively by salt creep, so that radionuclides should not be mobilized by the brine inflow. Finally, based on the information gathered and on geotechnical subcommittee's discussions, it appears that the geological setting of the WIPP is different from that of the Rhodes Yates oil field. The geological configuration near the WIPP site is likely to interfere with fluid movement thereby reducing the likelihood of flow from a hydraulic fracture into the repository. Therefore, in the committee's opinion, the Hartman Scenario is not likely to cause a problem in the performance of the repository.

2. *Intersection of a pressurized brine reservoir.* Groundwater containing high levels of dissolved solids (brine) may occur beneath the WIPP site either as discrete pockets (brine pockets) or as a saturated continuum. The committee uses the term "brine reservoir" to refer to both of these occurrences. At present, there is a great deal of uncertainty as to the location and form (i.e., discrete pocket or saturated continuum) of brine reservoirs beneath the WIPP repository. The committee recognizes that direct drilling through the repository into underlying high-pressure brine reservoirs could result in a release of radionuclides.

A survey study of brine reservoirs in the Castile Formation (Popielak et al., 1983) has suggested that the brine reservoirs in the area are not large enough to affect the safety of the WIPP site and that there is no high-pressure brine reservoir directly underlying the repository. However, this finding is challenged by Silva et al. (1999). Using data from test well WIPP-12, Silva demonstrated that the probability of a large brine reservoir, approximately 260 meters below the repository, is rather high. The issue remains unresolved at the present time. Direct drilling (see Appendix B, Boxes B.1, and B.4) into the WIPP repository would allow circulating drill fluid to bring radioactive materials to the surface through a borehole as cuttings or spallings. In the performance assessment, SNL evaluated different possibilities of drilling into a brine reservoir (see Appendix B).

In the committee's opinion, the upsurging pressure from drilling through a pressurized brine reservoir could be counteracted by the weight of drilling mud. However, the situation could be serious if the brine reservoir were large and contained a significant amount of energy. An intersection with such a reservoir, although extremely rare, could cause the well to blow out and could result in a catastrophic safety problem for the WIPP. In the committee's opinion, when the drillbit penetrates a brine reservoir below the repository, there would be an initial surge of brine flowing through the borehole into the repository, but the rate of brine inflow would decrease rapidly unless this high-pressure brine reservoir had a gas subpocket above it. Because of the low compressibility of brine, without a gas subpocket, the energy stored in the reservoir would not be sufficient to cause a large brine upsurge through the borehole into the repository.

It is therefore important to determine the existence of a brine reservoir directly below the repository. This would be done using seismic techniques, which cannot measure the pressure in the reservoir but can detect its size. The committee recognizes that small brine reservoirs, including brine occurring as a saturated continuum, could not be detected by seismic surveys, or other noninvasive remote sensing techniques. Most seismic surveys are performed from the surface. However, it is possible to perform measurements at a depth, such as in wells or from within the repository. There would be advantages to performing a seismic survey at repository depth (660 meters below the surface) because the unwanted signal from near-earth formations could be eliminated.

The committee is aware of the numerous geophysical surveys that have been performed on the WIPP area in the past (ETC, 1988; Popielak et al., 1983; Silva et al., 1999) and does not suggest repeating what has been already done. However, seismic interpretation technology has improved dramatically in the last decade. These improvements, including but not limited to the almost universal three-dimensional seismic techniques, have greatly enhanced resolution capability and are currently used in the oil industry. Detailed three-dimensional seismic studies results, however, are often highly proprietary because they are performed by the oil industry. The DOE could consider acquiring the results of these studies to obtain new information on possible brine reservoirs in the region.

In case a brine reservoir were found beneath the WIPP site and its size were larger than what is already taken into account in the PA, then the DOE should conduct an extensive review of the impact of such a reservoir on the repository performance. A basis would then exist to take appropriate action to ensure the safety of the repository. If the reservoir is pressurized, the option of drilling a well into it to release the pressure could be considered. In case of drilling, precautionary methods, such as directional drilling, should be taken to prevent brine from entering the repository.

Recommendation: **The committee recommends the utilization of seismic survey techniques to detect the presence of a large brine reservoir below the repository.[8] In case a brine reservoir were**

---

[8] The committee recognizes that small brine reservoirs, including brine occurring as a saturated continuum, could not be detected by seismic surveys, or other noninvasive remote sensing techniques.

found beneath the WIPP and its size were larger than what is already taken into account in the PA, then the DOE should conduct an extensive review of the impact of such a reservoir on the repository performance. A basis would then exist to take appropriate action to ensure the safety of the repository.

Mining Activities

A further human activity that could threaten the safety of the repository is potash mining in proximity of the WIPP site. Potash mining could impact the performance of the repository by modifying flow pathways in the overlying formations or by creating a path for brine intrusion, if methods such as flood or solution mining are employed. The potential impact of potash mining on WIPP performance is not considered significant, but it is important that the DOE monitor during the operational phase all mining activities in close proximity of the area addressed in the LWA to ensure that the WIPP repository performance is not affected.

After reviewing the analyses performed for the human intrusion scenarios as a part of the performance assessment and given the reasons mentioned above, the committee finds that oil, gas, and mineral activities will not unduly threaten the integrity of the repository. However, there are uncertainties associated with these extraction operations. These uncertainties could be reduced by monitoring and documenting oil, gas, and mineral activities. The DOE could establish a database on oil, gas, and mineral activities in the WIPP area containing information such as:

1. location, depth, and type of each well surrounding the WIPP site;
2. data on accidents or unusual events reported by drilling contractors or operators;
3. data on production-enhancing activities such as water or CO_2 flooding, hydraulic or cryogenic fracturing, and acidizing in surrounding wells;
4. production rates of oil, gas, and brine from nearby wells;
5. data on disposal of drill cuttings and brine from the operators;
6. data from abandoned wells, in particular those relevant to gas leakages; and
7. extent of potash mining in the vicinity of the LWA.

Recommendation: The committee recommends the development of a database to collect information on drilling, production enhancement, mining operations, well abandonments, and unusual events (accidents and natural events) in the vicinity of the WIPP site.

Baseline Radiogenic Analysis of Subsurface Fluids

The issue of baseline values for naturally occurring radioactive material (NORM) in the vicinity of the WIPP site is important for future monitoring of any changes in radioactivity levels in and around the site. The reason for concern is that subsurface oil and gas in the vicinity of the site already contains NORM. The potential discovery of radioactive material in oil and gas could mistakenly be assumed to come from the repository and thereby cast a doubt on the performance of the nearby WIPP.

One of the findings of the committee's interim report (Appendix A1) identified an absence of radiological baseline information for subsurface brines and hydrocarbons near the site, even though there has been extensive monitoring of radioactivity in the air, soils, fluvial sediments, surface water, shallow groundwater, and populace. Therefore, the committee recommended that the DOE develop and implement a plan to sample oil-field brines, petroleum, and solids associated with current hydrocarbon production to assess the magnitude and variability of naturally occurring radioactive material in the vicinity of the WIPP site. The radionuclides of interest include those that contribute to the site's NORM background radioactivity and those present in the TRU waste inventory destined for WIPP. The NORM activity may include contributions from potassium-40, isotopes of uranium and thorium, and daughter products such as isotopes of radium. Radionuclides in TRU waste include isotopes of uranium and TRU elements and, in remote-handled TRU waste, fission and activation products.

Since some TRU inventory radionuclides are not found commonly in nature, sampling to determine whether such radionuclides are present in the environment may be a good way to distinguish radioactivity due to NORM from that due to TRU waste. Further details can be found in Appendix A1. In its interim report, the committee recommended a simple but reliable analysis of the samples that do not include species depending on equilibria that can be shifted by a change in the chemical or physical parameters of the sample.

In response to the interim report, the DOE stated that the New Mexico State University Carlsbad Environmental Monitoring and Research Center (CEMRC) has undertaken a project to carry out the recommended assessment, as part of CEMRC's WIPP environmental monitoring project (Appendix A2). This project will include "completion of a database of active wells and operators, development of sample collection and handling plans, and identification of commercial sample collection services." The CEMRC has also developed analytical methods for NORM in subsurface fluids to complement standard methods. The committee supports and encourages the pursuit of this initiative.

Recommendation: **The committee recommends that the DOE continue the implementation of its plan to sample oil-field brines, petroleum, and solids associated with current and future hydrocarbon production, as necessary to assess the magnitude and variability of NORM in the vicinity of the WIPP site for baselining purposes.**[9]

---

[9] On March 12, 2001 the DOE-Carlsbad Field Office informed the committee that the efforts to collect data on NORM have received little support from oil companies and that cooperation seems unlikely. The small number of positive responses received would still not provide enough information to constitute a representative baseline of NORM in the region.

3

National Transuranic Waste Management Program

The National Transuranic Waste Management Program, also called the National TRU Program, addresses waste acceptance criteria and requirements for packaging and shipping waste to the WIPP repository. One of the committee's tasks is to identify areas for improvement in the National TRU Program that may increase safety to workers and the public, system throughput, efficiency, or cost-effectiveness. The National TRU Program was reviewed in detail in the committee's interim report (Appendix A1). This chapter gives a status report on the issues discussed in the interim report and reviews other issues that have emerged during the committee's deliberations. The issues addressed in this chapter relate to two areas: (1) *waste characterization and packaging* and (2) *waste transportation.*

WASTE CHARACTERIZATION AND PACKAGING

The committee has identified opportunities for improvement in the TRU waste management system concerning waste characterization and packaging requirements and the total inventory of organic material in the repository.

Waste Characterization and Packaging Requirements

The issues of waste characterization and packaging requirements have been discussed in detail in the interim report (Appendix A1). The principal finding was that many requirements and specifications concerning waste characterization and packaging lacked a safety or legal basis. In addition, many of these same requirements resulted in health and safety risks and added costs. The added safety risks derive from radiation exposure of workers due to the extra handling of waste imposed by some of the requirements. For instance, visual examination of a fraction of waste stream containers to confirm radiography results and information from the history of the container (acceptable knowledge) is a procedure not required by the EPA that increases radiation exposure of workers. The committee recommended in its interim report that the DOE eliminate self-imposed waste characterization requirements that lack a safety or legal basis.

The committee is encouraged by progress made since the interim report to eliminate unnecessary procedures. In particular, the DOE has initiated a program to review all waste characterization and packaging requirements and to reduce or eliminate those that do not contribute to improved safety or that are not required by law. The DOE may obtained a tenfold reduction of the number of containers to be opened for visual examination by requesting a modification of the WIPP's Hazardous Waste Facility Permit (Appendix A2).

Recommendation: **The committee recommends that the DOE's efforts to review waste characterization and packaging requirements continue and that changes be implemented over the entire National TRU Program. The committee recommends that the resources required to complete these improvements be made available by the DOE.**

Total Inventory of Organic Material in the Repository

A new issue concerning waste characterization has emerged since the committee visited the WIPP site in May 2000. This issue addresses the regulatory limits on the total inventory of organic material allowed in the repository. The performance assessment indicates that there could be significant carbon dioxide generation in the repository due to the decomposition of organic material. Although the committee does not consider gas generation an important safety issue (see Chapter 2), it is concerned whether the current monitoring program will provide the information required to assess compliance with total repository limits of organic material. Title 40 CFR 194.24 states that "the Department [of Energy] shall specify the limiting value ... of the total inventory of such waste proposed for disposal" (EPA, 1996). The DOE has therefore established the limit for organic material in the repository to be 20 million kilograms (DOE, 1996; Table 4-0) on the basis of the average waste composition.

However, the DOE's definition of "waste" includes only what is inside the waste container and does not include either the container itself or any of the auxiliary material buried with the waste. Examples of such auxiliary material are plastic films used to stabilize drums for shipping and handling, plastic bags and corrugated cardboard used as magnesium oxide containers, wooden waste boxes, plastic liners of waste drums, and pressed wood "slip sheets" used between layers of drums and waste boxes. Figure 3.1 shows a picture of waste and auxiliary material in one of the rooms of the repository. Thus, there is a considerable inventory of materials, mostly cellulosics, that are not considered TRU waste but are foreign to the natural setting of the Salado Formation. The principal concern of the committee is that the auxiliary material does not appear to be accurately inventoried. Therefore, it is impossible to know whether the total organic material limit is exceeded.

Recommendation: **The committee recommends a risk-based analysis of the total organic material regulatory limits in the WIPP. If accounting for the organic material is important to the safety of the repository, an inventory record system should be implemented as soon as possible to provide a basis for meaningful safety analysis.**

WASTE TRANSPORTATION

The committee has examined various aspects of the WIPP TRU waste transportation system, focusing on system safety and on cost-effectiveness of planned and ongoing activities. In its interim report (Appendix A1), the committee reviewed the DOE's TRANSportation tracking and COMmunication (TRANSCOM) system and its emergency response program. Two other issues have been revisited in this report: the potential use of rail as a shipping option for a fraction of TRU waste and the gas genera-

Figure 3.1 Standard waste boxes and packs stacked in one room of the WIPP repository. Notice the layers of plastic film around the drums. SOURCE: DOE, 2000d.

tion safety analysis for the Transuranic Package Transporter, Model II (TRUPACT-II) containers. Figure 3.2 shows the internal structure of a TRUPACT-II container. Figure 3.3 shows a truck transporting three TRUPACT-II containers.

DOE's Communication and Notification Program

In its interim report (Appendix A1), the committee recommended that the DOE improve the reliability and ease of use of the TRANSCOM system. On November 21, 2000, a truck hauling waste to WIPP strayed from its designated route as the driver missed the exit from Interstate 25 onto Route 285 toward Carlsbad. The driver proceeded 27 miles before the New Mexico State Police, equipped with a TRANSCOM system, realized the error and turned the truck around. It appears that the TRANSCOM headquarters situated in Oak Ridge, Tennessee, did not notify the driver until a state policeman noticed the error (DOE, 2001). This "strayed truck" episode is an example of the poor reliability of the system, not from a technical point of view since the TRANSCOM was apparently functioning correctly, but from the perspective of the human factor.

The DOE appears to be systematic and expeditious in its development and use of a new, efficient, comprehensive, and state-of-the-art communication and notification system, known as TRANSCOM 2000. The new system will use off-the-shelf, advanced information and communication technologies to track shipments from start to end. Full-scale implementation of TRANSCOM 2000 is scheduled for June 2001. After discussion with transportation management staff and in reaction to the DOE's response to its

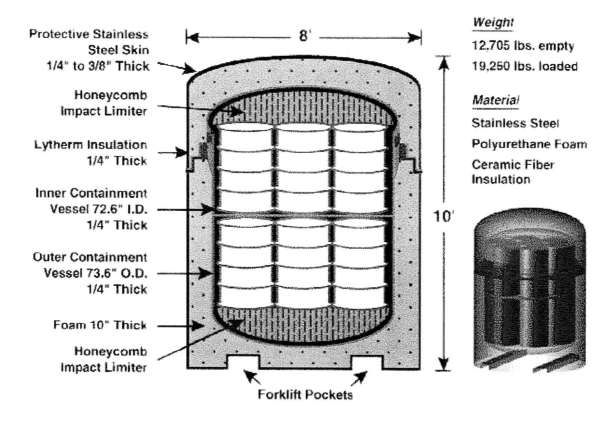

Figure 3.2 Structure of TRUPACT-II container, certified by the USNRC. SOURCE: DOE, 2000l.

Figure 3.3 Truck transporting three TRUPACT-II containers to the WIPP. SOURCE: DOE, 2000m.

interim report (Appendix A2), the committee finds that overall, the DOE has taken active steps to address concerns about the reliability and ease of use of the TRANSCOM system. Moreover, the DOE has integrated new features into TRANSCOM 2000, such as alarms, more frequent satellite and computer tracking, and stronger training for truck drivers to avoid future "strayed truck" episodes. Other suggestions to improve the safety and reliability of TRANSCOM could be the use of checklists, key schedule reporting, and "call-ins" at important route changes.

There may also be further opportunities to improve the performance of the TRANSCOM 2000 system. For instance, integrating TRANSCOM 2000 with other corridor states' information technology programs such as intelligent transportation systems (ITS). More generally, TRANSCOM 2000 must meet performance-monitoring standards similar to those of other advanced systems, such as the air traffic control system, particularly as shipments become routine and drivers and supervisors may become complacent. Investment in tracking and communication systems for the WIPP will also be useful for future radioactive waste transportation systems.

Recommendation: **The committee recommends that the DOE implement as soon as possible the new TRANSCOM 2000 communication and notification system. Moreover, because human factors are an important element of transportation system quality, TRANSCOM 2000 should include methods to minimize the occurrence and impact of human errors.**

DOE's Emergency Response Program

Concerning the emergency response program for the WIPP, the committee recommended, in its interim report (Appendix A1), that the DOE explore with corridor states and other interested parties how to develop processes and tools for maintaining up-to-date spatial information on the location, capabilities, and contact information for the following:

- responders,
- medical facilities,
- recovery equipment,
- regional response teams, and
- other resources that might be needed to support effective emergency response in the event of a transportation incident involving a WIPP shipment.

This recommendation was made in recognition of the fact that, presently, there is no quality control program in existence to evaluate periodically and systematically the extent of training, emergency capabilities, and deficiencies within the states and along WIPP transportation corridors. The committee fully understands and recognizes that the primary responsibility for management and response to hazardous material incidents in transportation rests with state and local authorities and jurisdictions. Although WIPP corridor states actively coordinate in varying degrees with the DOE to ensure the safety of WIPP shipments, the general public may often view this responsibility as ultimately resting with the DOE as the system manager. The public might well expect qualified and trained emergency response coverage along an entire route. In the committee's view, the DOE could face heavy criticism if an event demonstrates weaknesses in the emergency response program, regardless of whether the safety consequences are serious. Any system-level integration necessary to ensure adequate emergency response would have to recognize and coordinate among the jurisdictional boundaries of the various responsible state and local agencies.

To date, the committee is concerned about the progress being made in the emergency response area. For example, only 7 of the 20 states situated along the transportation corridor participated in the last DOE emergency response training class in Carlsbad, New Mexico (DOE, 2000e). The committee acknowledges the challenges faced by the respective states in providing resources to ensure adequate coverage. There continues to be a need for the DOE to facilitate the involvement of states and other interested parties to determine where emergency response capabilities are lacking along transportation routes and to support the states in correcting deficiencies. The committee is encouraged by the new DOE training program, through which DOE trainers have traveled to Indiana, Colorado, Louisiana, and Nevada to teach emergency response professionals what to do in case of an accident involving a WIPP shipment (Westinghouse News, 2001a,b,c,d). A further example for DOE to improve the corridor states' involvement in the emergency response program is to organize training courses through distant learning.

Recommendation: **The committee recommends that the DOE facilitate the involvement of states in developing and maintaining an up-to-date, practical, and cost-effective spatial information database system to coordinate emergency responses. The DOE should also develop an ongoing assessment program for states' emergency response capabilities and allocate training resources to address deficiencies in coverage along WIPP routes.**

Rail as a Transportation Option for Certain TRU Waste

In its interim report (Appendix A1), the committee recommended that DOE reduce the number of truckloads required to transport waste to WIPP, thereby reducing the associated transportation risks.[1] The committee suggested that a way to reduce the number of shipments is to reevaluate the technical and regulatory feasibility of shipping high-wattage TRU waste using a railcar shipping system. The WIPP has access already to rail via a rail spur siding, which runs into the facility.

In response to this recommendation, the DOE (2000c) recently issued a report *CH-TRU Waste Transportation System Rail Study*. This study examined the feasibility of shipping CH-TRU waste from four DOE facilities to WIPP by commercial rail and compared the relative costs of using rail rather than the present use of the highway. The study also examined the feasibility and cost-effectiveness of using several alternative packaging to TRUPACT-II. TRUPACT-II containers, because of their size, shape, or regulatory limits, are not always efficiently utilized during transportation; therefore, an increased number of shipments or repackaging of the waste is sometimes required. The DOE concluded that rail shipment of TRU waste to WIPP might be competitive if certain conditions are satisfied. Those conditions involve negotiation of a more favorable rail rate and development of an alternate type B overpack to TRUPACT-II that would accommodate more packages, thus reducing the number of shipments required.

A recent article (Neill and Neill, 2000) asserts that rail offers considerable advantages, at least with respect to shipments from the Hanford and the Idaho National Engineering and Environmental Laboratory sites. The authors make specific recommendations concerning the use of rail that might enable the DOE to ship TRU waste more efficiently while reducing transportation risk. The committee suggests

---

[1] On November 2, 2000, a new type of container, called HalfPACT, designed to supplement TRUPACT-II for road transportation, was certified by the U.S. Nuclear Regulatory Commission. The new container is approximately 30 inches shorter than TRUPACT-II and can be utilized more efficiently to transport TRU waste. The DOE estimated that the new HalfPACT container will eliminate about 2,000 projected shipments to the WIPP site.

> **Sidebar 3.1 The ATMX Railcar System as an Alternative Transportation System?**
>
> "ATMX" is an acronym to denote the railcars used by the DOE to ship nuclear weapons components and TRU waste. "AT" stands for Atchison Topeka, the rail carrier. "M" signifies munitions, and "X" on a railcar signifies private ownership (in this case, by the U.S. government), rather than ownership by the railroad company. This system was used by the DOE (and formerly the U.S. Atomic Energy Commission) from about 1968 to 1989 to safely transport more than 1,100 shipments of CH-TRU waste from the Mound Laboratory and Rocky Flats to the Idaho National Engineering Laboratory. The ATMX (600 series) is a specially designed steel railcar with a bolted-on steel cover and an interior compartmentalized by steel frames. Closed steel boxes or bins are positioned and stored in each compartment, and internal packagings are placed in the boxes or bins. Internal packagings need only meet U.S. Department of Transportation (DOT) Type A package test standards and are relieved from Type B (accident-resistant) package test parameters. Each ATMX railcar can accommodate a maximum of 20 crates or 140 55-gallon steel drums. In June 1999, the DOT issued the tenth revision of DOT-E 5948, authorizing the shipment of TRU waste by rail from Miamisburg, Ohio (the Mound Laboratory), to a yet-to-be-designated DOE facility where it will be processed for eventual shipment to the WIPP in TRUPACT-II containers. This option appears to be a very reasonable and cost-effective method of transferring the relatively small amount of TRU waste at Mound to another DOE facility for processing as an alternative to setting up a facility at Mound itself. Since the ATMX system is not certified by the USNRC, its use for rail shipments directly to the WIPP is precluded by the provisions of the Land Withdrawal Act and the Agreement with the State of New Mexico, which require that shipments to the WIPP be in USNRC-certified packages. To obtain USNRC "approval," the DOE would have to support an application to the USNRC for exemption from certain test requirements for the Type B package mentioned in Title 10 CFR Part 71. For certain materials that eventually will not be transportable in TRUPACT-II containers due to high thermal loading, this would appear to be a desirable option for future consideration and possible pursuit by the DOE.

that the DOE develop a strategy to negotiate and reduce the overall rail freight costs and to identify the infrastructure (e.g., costs, emergency preparedness, and schedules) necessary for rail shipments.

In its interim report, the committee recommended the ATMX railcar system as an alternative transportation system for certain materials (see Sidebar 3.1). Specifically, the committee recommended that a risk-informed study should be prepared by the DOE to support an application to the United States Nuclear Regulatory Commission (USNRC) authorizing the use of the ATMX railcar system for shipments of inner waste packages that are unsuitable for placement in TRUPACT-II containers.

Recommendation: **The committee recommends that all reasonable transportation options including reduction in the number of shipments, such as rail and road transportation with better-adapted containers, should be part of the decision-making process of transporting TRU waste from generator and storage sites to the WIPP. Future transportation studies should consider railway shipments and their impact on both the safety and the cost of the program. The DOE should also** continue to pursue the development of packaging alternatives for materials not suitable for TRUPACT-II containers.

Gas Generation Safety Analysis for TRUPACT-II Containers

The issue of hazardous gas generation in TRUPACT-II shipping containers stems from a U.S. Nuclear Regulatory Commission requirement (USNRC, 1999). The requirement states that, for the shipping container, "hydrogen and other flammable gases comprise less than 5% by volume of the total gas inventory within any confinement volume." The problem is whether a flammable mixture could be generated and trigger an ignition, exothermic reaction, or explosive event with sufficient energy to breach the containment. Excessively restrictive gas generation requirements have severe consequences. The waste is repackaged to redistribute waste in containers to meet the wattage limits derived from gas generation requirements. This repackaging of waste exposes workers to radiation and increases the number of containers, thereby diluting the waste into a greater volume. Transportation-related risks (and costs) are also incurred in repackaging because the extra containers require additional shipping loads and additional truck trips, thereby increasing the likelihood of accidents.

For instance, plutonium-238 found in CH waste is considered a "high-wattage waste" because of its high specific activity (17.3 curies per gram). The USNRC significantly restricts the amount of plutonium-238 that can be transported by TRUPACT-II because of gas generation concerns. The DOE estimates that the repackaging of plutonium-238 in CH waste may involve more than a tenfold increase in the number of shipments of plutonium-238, to as many as 150,000 extra drums (Lechel and Leigh, 1998). The USNRC uses a decay heat limit in watts, originally established by the DOE, based on limiting the volume of hydrogen to less than 50 milliliters per liter of volume in the "innermost confinement barrier." According to waste acceptance criteria for the WIPP, the wattage limit for TRUPACT-II containers is 40 watts (DOE, 1999).

Thus, there is also the matter of what constitutes the innermost confinement barrier in TRUPACT-II, since the containers may consist of separate individual plastic bags of waste (see Figure 3.2). One interpretation is that the requirement applies to these "inner packages." Obviously, there are situations in which such an interpretation would make the flammable gas volume limitation a severe constraint on TRU waste shipments, given the plastic bag packaging practice and the number of different sizes that may occur in a single TRUPACT-II container. Finally, it is the understanding of the committee that the 5 percent volume limitation on hydrogen is intended to preclude the need for a specific safety analysis, which suggests that this limitation is a major source of conservatism and may not be cost-effective or risk-informed.

The committee was unable to verify the technical basis for the several sub-issues that are involved, including a realistic assessment of the conditions that could result in an explosive event in TRUPACT-II containers and a clear definition of what constitutes the innermost barrier. As already recommended in its interim report, the committee reiterates that a risk-informed analysis of WIPP-specific shipments would contribute to a better understanding of the real safety issues and, perhaps, provide a basis for alternative cost-effective criteria while reducing the risk.

In its response to the interim report (Appendix A2), the DOE agreed with the committee's recommendation that a safety analysis be performed to determine the quantity of hydrogen that, upon ignition, could damage the TRUPACT-II shipping container and possibly rupture the seals of the package. The committee is aware of and supports the DOE's initiative of obtaining more realistic G-values[2] for hydro-

---

[2] The G-value is the measure of radiolytic yield. It is expressed by the number of molecules, in this case of H_2, produced by 100 electronvolts of the ionizing radiation's energy absorbed by the medium, in this case the TRU waste.

gen generation and of exploring the use of hydrogen getters and inerted inner containers as a means of increasing wattage limits for transportation. Moreover, the DOE has applied for a revision of the USNRC's certificate of compliance to authorize the use of lower G-values for hydrogen generation based on matrix depletion, options for mixing of shipping categories, and use more realistic G-values for non-gas generating materials that are present. The committee supports this request.

Recommendation: **The committee recommends a risk-informed analysis of WIPP specific shipment issues to identify core problems related to hydrogen generation and, perhaps, provide a basis for alternative cost-effective criteria while reducing the risk. The committee recommends the use of such risk-informed analysis in the application for revision of the USNRC certificate of compliance concerning hydrogen generation limits for transportation purposes.**

4

Summary

The committee is confident that the WIPP can meet its general performance objectives as requested by the certification process. However, uncertainties remain in the long-term performance of the repository. Some of the recommendations in this report were released in the committee's interim report (Appendix A1) to which the DOE has responded with a number of actions taken (see Appendix A2). The committee encourages implementation of the improvements suggested by the DOE to address its recommendations. In Chapters 2 and 3, the committee addresses some new issues concerning the operation and long-term safety of the WIPP and reiterates for emphasis some of the recommendations of the interim report. This chapter closes the study with an overarching finding and recommendation.

OVERARCHING FINDING

The committee finds that the monitoring of selected performance indicators during the estimated 35-year or longer pre-closure phase of the WIPP is needed to possibly enhance confidence in the long-term safety performance of the repository. Although 35 to possibly 100 years is a short time compared to the 10,000-year period of compliance, the committee believes that it is long enough to reduce the uncertainties in many critical performance parameters. The rates of important processes such as salt creep, brine inflow (if any), and gas generation are predicted to be highest during this period; therefore, monitoring during the pre-closure phase is particularly important. Moreover, the committee finds that there are a number of specific actions that can be taken in the National TRU Program to facilitate operation of the WIPP while increasing safety and reducing costs.

OVERARCHING RECOMMENDATION

The committee recommends that the DOE develop and implement a program during the pre-closure phase to monitor selected performance indicators that specifically relate to the creation of a radionuclide source term and to pathways for radionuclide transport. Monitoring should continue throughout the pre-closure phase and longer, if possible. Emphasis in the monitoring should be on waste mobilization and transport mechanisms, including brine inflow, gas generation, geochemical reactions, room sealing, and surface and subsurface hydrology. The committee recommends that the results of the on-site monitoring program be used to improve the performance assessment for recertification purposes. These results will determine whether the need for a new performance assessment is warranted. Given the uncertainties, it is impossible to predict if the results of the monitoring program will be different than those modeled by the performance assessment. However, it is important to ensure that, if there are changes, these will be detected.[1] Moreover, actions should be taken to improve and better define the National TRU Program for issues related to waste characterization and packaging requirements, total inventory of organic materials, communication and notification system, emergency response training, and gas generation during transportation.

The committee did not have all of the information necessary to prioritize the issues mentioned in this report. However, it has provided a selected number of recommendations that are believed to improve the operation and long-term safety of the WIPP. The committee recognizes that the recommendations in this report will have some economic impact on the transuranic waste management program. The DOE needs to balance costs against the improved assurance of facility performance in the longer term.

---

[1] Only measurable changes are important to verify the performance of the repository; for instance, a few drops of brine do not imply that the repository is not in compliance with containment requirements.

References

Apostolakis, G. E., C. Guedes Soares, S. Kondos, J. C. Helton, and M. G. Marietta, 2000. The 1996 Performance Assessment for the Waste Isolation Pilot Plant. Reliability Engineering and System Safety. Special Issue: The 1996 Performance Assessment for the Waste Isolation Pilot Plant; 69(1-3):1-456.

Berglund, J. W., J. Myers, L. R. Lenke, 1996. Memorandum to Margaret Chu. Estimate of the Tensile Strength of Degraded Waste for use in Solids Blowout. July 19, 1996. Albuquerque, N.M.: New Mexico Engineering Research Institute.

Berglund, J. W., J. W. Garner, J. C. Helton, J. D. Johnson, and L. N. Smith, 2000. Direct Releases to the Surface and Associated Complementary Cumulative Distribution Functions in the 1996 Performance Assessment for the Waste Isolation Pilot Plant: Cuttings, Cavings and Spallings. Reliability Engineering and System Safety. Special Issue: The 1996 Performance Assessment for the Waste Isolation Pilot Plant; 69(1-3):263-304.

Bredehoeft, J., and J. Gerstle, 1997. The Hartman Scenario Revisited, Implications for WIPP. Prepared for New Mexico Attorney General, August, 1997. La Honda, Calif.: The Hydrodynamics Group.

Callahan, G. D., and K. L. Devries, 1991. Analyses of Backfilled Transuranic Wastes Disposal Rooms. SAND91-7051. Albuquerque, N.M.: Sandia National Laboratories.

Corbet, T. F., and P. N. Swift, 2000. The Geologic and Hydrogeologic Setting of the Waste Isolation Pilot Plant. Reliability Engineering and System Safety. Special Issue: The 1996 Performance Assessment for the Waste Isolation Pilot Plant; 69(1-3):47-58.

Earth Technology Corporation (ETC), 1988. Final Report for Time Domain Electromagnetic (TDEM) Surveys at the WIPP Site. SAND87-7144. Albuquerque, N.M.: Sandia National Laboratories.

Francis, A. J., J. B. Gillow, and M. R. Giles, 1997. Microbial Gas Generation Under Expected Waste Isolation Pilot Plant Repository Conditions. SAND 96-2582. Albuquerque, N.M.: Sandia National Laboratories.

Galson, D. A., P. N. Swift, D. R. Anderson, D. G. Bennett, M. B. Crawford, T. W. Hicks, R. D. Wilmot, and G. Basabilvazo, 2000. Scenario Development for the Waste Isolation Pilot Plant Compliance Certification Application. Reliability Engineering and System Safety. Special Issue: The 1996 Performance Assessment for the Waste Isolation Pilot Plant; 69(1-3):129-150.

Gerstle, W., and J. Bredehoeft, 1997. Linear Elastic Model for Hydrofracture at WIPP and Comparison with BRAGFLO Results. September, 1997. Prepared for the New Mexico Attorney General, Department of Civil Engineering. Albuquerque, N.M: University of New Mexico.

Hansen, F. D., and M. K. Knowles, 2000. Design and Analysis of a Shaft Seal System for the Waste Isolation Pilot Plant. Reliability Engineering and System Safety. Special Issue: The 1996 Performance Assessment for the Waste Isolation Pilot Plant; 69(1-3):87-98.

Hansen, F. D., M. K. Knowles, T.W. Thompson, M. Gross, J. D. McLennan, and J. F. Schatz, 1997. Description and Evaluation of a Mechanistically Based Conceptual Model for Spall. SAND97-1369. Albuquerque, N.M.: Sandia National Laboratories.

REFERENCES

Helton, J. C., D. R. Anderson, G. Basabilvazo, H.-N. Jow, and M. G. Marietta, 2000a. Conceptual Structure of the 1996 Performance Assessment for the Waste Isolation Pilot Plant. Reliability Engineering and System Safety. Special Issue: The 1996 Performance Assessment for the Waste Isolation Pilot Plant; 69(1-3):151-166.

Helton, J. C., M.-A. Martell, and M. S. Tierney, 2000b. Characterization of Subjective Uncertainty in the 1996 Performance Assessment for the Waste Isolation Pilot Plant. Reliability Engineering and System Safety. Special Issue: The 1996 Performance Assessment for the Waste Isolation Pilot Plant; 69(1-3):191-204.

Helton, J. C., F. J. Davis, and J. D. Johnson, 2000c. Characterization of Stochastic Uncertainty in the 1996 Performance Assessment for the Waste Isolation Pilot Plant. Reliability Engineering and System Safety. Special Issue: The 1996 Performance Assessment for the Waste Isolation Pilot Plant; 69(1-3):167-190.

Helton, J. C., J. E. Bean, K. Economy, J. W. Garner, R. J. MacKinnon, J. Miller, J. D. Schreiber, and P. Vaughn, 2000d. Uncertainty and Sensitivity Analysis for the Two-Phase Flow in the Vicinity of the Repository in the 1996 Performance Assessment for the Waste Isolation Pilot Plant: Undisturbed Conditions. Reliability Engineering and System Safety. Special Issue: The 1996 Performance Assessment for the Waste Isolation Pilot Plant; 69(1-3): 227-262.

Helton, J. C., J. E. Bean, K. Economy, J. W. Garner, R. J. MacKinnon, J. Miller, J. D. Schreiber, and P. Vaughn, 2000e. Uncertainty and Sensitivity Analysis for the Two-Phase Flow in the Vicinity of the Repository in the 1996 Performance Assessment for the Waste Isolation Pilot Plant: Disturbed Conditions. Reliability Engineering and System Safety. Special Issue: The 1996 Performance Assessment for the Waste Isolation Pilot Plant; 69(1-3):263-304.

Howard, B. A., M. B. Crawford, D. A. Galson, and M. G. Marietta, 2000. Regulatory Basis for the Waste Isolation Pilot Plant Performance Assessment. Reliability Engineering and System Safety. Special Issue: The 1996 Performance Assessment for the Waste Isolation Pilot Plant; 69(1-3):109-128.

Idaho National Engineering and Environmental Laboratory (INEEL). 1998. TRUPACT-II Matrix Depletion Program. Final Report. INEEL/EXT-98-00987. Rev. 0. Idaho Falls: Idaho National Engineering and Environmental Laboratory.

Jensen, A. L., R. L. Jones, E. N. Lorusso, and C. L. Howard, 1993. Large-Scale Brine Inflow Data Report for Room Q Prior to November 25, 1991. SAND92-1173. Albuquerque, N.M.: Sandia National Laboratories.

Knowles, M. K., and K. M. Economy, 2000. Evaluation of Brine Inflow at a Waste Isolation Pilot Plant. Water Environment Research; 72(4):397-404.

Knowles, M. D., F. D. Hansen, T. W. Thompson, J. F. Schatz, and M. Gross, 2000. Review and Perspectives on Spallings Release Models in the 1996 Performance Assessment for the Waste Isolation Pilot Plant. Reliability Engineering and System Safety. Special Issue: The 1996 Performance Assessment for the Waste Isolation Pilot Plant; 69(1-3):331-341.

Krumhansl J. L., M. A. Molecke, H. W. Papenguth, and L. H. Brush, 1999. A Historical Review of Waste Isolation Pilot Plant Backfill Development, Proc. ICEM '99 Nagoya, Japan.

Lappin, A. R., and R. L. Hunter, eds., 1989. Systems Analysis, Long-Term Radionuclide Transport, and Dose Assessments, Waste Isolation Pilot Plant (WIPP), Southeastern New Mexico. SAND89-0462. Albuquerque, N.M.: Sandia National Laboratories.

Larson, K. W, 2000. Development of the Conceptual Models for Chemical Conditions and Hydrology Used in the 1996 Performance Assessment for the Waste Isolation Pilot Plant. Reliability Engineering and System Safety. Special Issue: The 1996 Performance Assessment for the Waste Isolation Pilot Plant; 69(1-3):59-86.

Lechel, D. J., and C. D. Leigh, 1998. Plutonium-238 Transuranic Waste Decision Analysis. SAND98-2629. Albuquerque, N. M.: Sandia National Laboratories.

Molecke, M. A., 1979a. Gas Generation from Transuranic Waste Degradation: Data Summary and Interpretation. SAND79-1245. Albuquerque, N.M.: Sandia National Laboratories.

Molecke, M. A., 1979b. Gas Generation Potential from TRU Wastes. Chapter 3: Summary of Research and Development Activities in Support of Waste Acceptance Criteria for WIPP. SAND79-1305. Albuquerque, N.M.: Sandia National Laboratories.

National Research Council (NRC), 1957. The Disposal of Radioactive Waste on Land. Washington, D.C.: National Academy Press.

NRC, 1984. Review of the Scientific and Technical Criteria for the Waste Isolation Pilot Plant (WIPP). Washington, D.C.: National Research Council.

NRC, 1988. Report on Brine Accumulation in the WIPP Facility. Washington, D.C.: National Research Council.

NRC, 1992. WIPP Letter Report (addressing the underground experimental plan with TRU wastes) to the Honorable Leo P. Duffy, Assistant Secretary of the U.S. DOE Office of Environmental Restoration and Waste Management. Washington, D.C.: National Academy Press.

NRC, 1996a. The Waste Isolation Pilot Plant, A Potential Solution for the Disposal of Transuranic Waste. Washington, D.C.: National Academy Press.

NRC, 1996b. Rock Fractures and Fluid Flow. Washington, D.C.: National Academy Press.

NRC, 2000. Improving Operations and Long-Term Safety of the Waste Isolation Pilot Plant, Interim Report. Washington, D.C.: National Academy Press.

Neill, H. R., and R. H. Neill, 2000. Transportation of Transuranic Nuclear Waste to WIPP: A Reconsideration of Truck Versus Rail for Two Sites. Natural Resources Journal; 40(1):93-124.

Novak C. F., and R. C.Moore,1996. Estimates of dissolved concentrations for +III, +IV, +V, and +VI actinides in a Salado and a Castile brine under anticipated repository conditions. Memorandum to M. D. Siegel, March 28, 1996. Albuquerque, N.M.: Sandia National Laboratories.

Oversby, V. M., 2000. Plutonium Chemistry Under Conditions Relevant for WIPP Performance Assessment. Review of Experimental Results and Recommendations for Future Work. EEG-77, DOE/AL58309-77. Albuquerque, N.M.: Environmental Evaluation Group.

Papenguth, H., J. Kelly, D. Lucero, J. Krumhansl, H. Anderson, P. Zhang, F. Stohl, N. Brodsky and D. Coffey, 1999. MgO Disposal Room Chemistry. Not published. Presentation at Sandia National Laboratories Technical Exchange, February 10, 1999, Carlsbad, New Mexico.

Popielak, R. S., R. L Beauheim, S. R. Black, W. E. Coons, C. T. Ellingson, and R. L. Olsen, 1983. Brine Reservoirs in the Castile Formation, Waste Isolation Pilot Plant (WIPP) Project, Southeastern New Mexico. TME-3153. Carlsbad, N.M.: U.S. Department of Energy.

Ramsey, J. L., R. Blaine, J. W. Garner, J. C. Helton, J. D. Johnson, L. N. Smith and M. Wallace. 2000. Radionuclide and Colloid Transport in the Culebra Dolomite and Associated Complementary Cumulative Distribution Functions in the 1996 Performance Assessment for the Waste Isolation Pilot Plant. Reliability Engineering and System Safety. Special Issue: The 1996 Performance Assessment for the Waste Isolation Pilot Plant; 69(1-3): 397-420.

Rechard, R. P., 2000. Historical Background on Performance Assessment for the Waste Isolation Pilot Plant. Reliability Engineering and System Safety. Special Issue: The 1996 Performance Assessment for the Waste Isolation Pilot Plant; 69(1-3):5-46.

Silva, M. K., 1996. Fluid Injection for Salt Water Disposal and Enhanced Oil Recovery as a Potential Problem for the WIPP: Proceedings of a June 1995 Workshop and Analysis. EEG-62. Albuquerque, N.M.: Environmental Evaluation Group.

Silva, M. K., D. F. Rucker, and L. Chaturvedi, 1999. Resolution of the Long-Term Performance Issues at the Waste Isolation Pilot Plant. Rick Analysis; 19(2):1003-1016.

Stoelzel, D. M and D. G.O'Brien, 1996. Analysis Package for the BRAGFLO Direct Release Calculations (Task 4) of the Performance Assessment Analysis Supporting the Compliance Certification Application. Sandia WIPP Central Files, WPO #45020. Albuquerque, N.M.: Sandia National Laboratories.

Stoelzel, D. M, D. G. O'Brien, J. W. Garner, J. C. Helton, J. D. Johnson, and L. N. Smith, 2000. Direct Releases to the Surface and Associated Complementary Cumulative Distribution Functions in the 1996 Performance Assessment for the Waste Isolation Pilot Plant: Direct Brine Release. Reliability Engineering and System Safety. Special Issue: The 1996 Performance Assessment for the Waste Isolation Pilot Plant. 69(1-3):343-368.

Stone, C. M., 1997. Final Disposal Room Response Calculations. SAND97-0795. Albuquerque, N.M.: Sandia National Laboratories.

Swift, P. N., R. L. Beauheim, P. Vaughn, and K. W. Larson, 1997. Response to John Bredehoeft's memorandum of July 1997 titled "Rebuttal: Technical Review of the Hartman Scenario: Implication for WIPP by Swift, Stoelzel, Beauheim, and Vaughn." Memorandum to Marget S. Y. Chu, August 1997. Albuquerque, N.M.: Sandia National Laboratories.

Telander, M. R., and R. E. Westerman, 1997. Hydrogen Generation by Metal Corrosion in Simulated Waste Isolation Pilot Plant Environments. SAND96-2538. Albuquerque, N.M.: Sandia National Laboratories.

U.S. Congress, 1992. Waste Isolation Pilot Plant Land Withdrawal Act. P.L. 102-579. Legislative Report for the 102nd Congress.

U.S. Department of Energy (DOE), 1996a. Title 40 CFR Part 191 Compliance Certification Application for the Waste Isolation Pilot Plant (21 volumes). DOE/CAO 1996-2184. Carlsbad, N.M: Carlsbad Area Office.

DOE, 1996b. Citizens' Guide to the Waste Isolation Pilot Plant Compliance Certification Application to the EPA. DOE/CAO 1996-1207. Carlsbad, N.M: Carlsbad Area Office

DOE, 1999. DOE Waste Acceptance Criteria for the Waste Isolation Pilot Plant. Revision 7. November 8, 1999. DOE/WIPP-069. Carlsbad, N.M: Carlsbad Field Office.

DOE, 2000a. National TRU Waste Management Plan. DOE/NTP-96-1204, Draft, July 31, 2000. Revision 2. Carlsbad, N.M.: Carlsbad Area Office.

DOE, 2000b. Re-evaluation of MgO Emplacement Doses at WIPP. WIPP Radiological Control Position Paper. 2000-07. Draft. Carlsbad, N.M.: Carlsbad Area Office.

REFERENCES

DOE, 2000c. CH-TRU Waste Transportation System Rail Study. DOE/WIPP 00-2016. Carlsbad, N.M.: Carlsbad Area Office.

DOE, 2000d. The Waste Isolation Pilot Plant. Pioneering Nuclear Waste Disposal. DOE/CAO-00-3124. Carlsbad, N.M.: Carlsbad Area Office.

DOE, 2000e. Emergency Response Training Draws Professionals from Seven States. In United States Department of Energy News, August 18, 2000 [on-line]. Available: http://www.wipp.carlsbad.nm.us/pr/2000-1/trainer.pdf [December 15, 2000].

DOE, 2000f. In Compliance Certification Application for the Waste Isolation Pilot Plant. Citizen's Guide [on-line]. Available: http://www.wipp.carlsbad.nm.us/library/cca/cca.htm [November 30, 2000]

DOE, 2000g. Location of the WIPP site. [on-line]. Available: http://www.wipp.carlsbad.nm.us/photo&graphics/map1.htm [November 30, 2000].

DOE, 2000h. The WIPP Facility and Stratigraphic Sequence [on-line]. Available: http://www.wipp.carlsbad.nm.us/photo&graphics/graph2.htm [November 30, 2000].

DOE, 2000i. Picture of Typical Transuranic Waste [on-line]. Available: http://www.wipp.carlsbad.nm.us/photo&graphics/tranwst.htm [November 30, 2000].

DOE, 2000j. Defense Transuranic Waste Generating and Storage Sites [on-line]. Available: http://www.wipp.carlsbad.nm.us/routes.htm [November 30, 2000].

DOE, 2000k. "Figure 6-38 and Figure 6-35." In Compliance Certification Application for the Waste Isolation Pilot Plant [on-line]. Available: http://www.wipp.carlsbad.nm.us/library/cca/cca.htm [2000, November 30].

DOE, 2000l. Structure of TRUPACT-II container, certified by the USNRC [on-line]. Available: http://www.wipp.carlsbad.nm.us/photo&graphics/graph4txt.htm [December 1, 2000].

DOE, 2000m. Truck Transporting three TRUPACT-II containers to the WIPP [on-line]. Available: http://www.wipp.carlsbad.nm.us/photo&graphics/graph4txt.htm [December 1, 2000].

DOE, 2001. DOE News. Tri-State Motor Transit to Resume Shipping Transuranic Waste to WIPP. January 19, 2001. [on-line]. Available: http://www.wipp.carlsbad.nm.us/pr/2001/TRIState.pdf [February 12, 2001].

U.S. Environmental Protection Agency (EPA), 1985. 40 CFR Part 191: Environmental Standards for the Management and Disposal of Spent Nuclear Fuel, High-Level and Transuranic Radioactive Wastes: Final Rule. September 19, 1985. Federal Register; 50(182):38066-38089.

EPA, 1993. 40 CFR Part 191: Environmental Radiation Protection Standards for the Management and Disposal of Spent Nuclear Fuel, High-Level and Transuranic Radioactive Waste. December 20, 1993. Federal Register; 58(242):66398-66416.

EPA, 1996. 40 CFR Part 194: Criteria for the Certification and Re-Certification of the Waste Isolation Pilot Plant's Compliance with the 40 CFR Part 191 Disposal Regulations: Final Rule. February 9, 1996. Federal Register; 61(28): 5224-5245.

EPA, 1998. 40 CFR Part 194: Criteria for the Certification and Re-certification of the Waste Isolation Pilot Plant's Compliance with the 40 CFR Part 191 Disposal Regulations: Certification Decision. Final Rule. May 18, 1998. Federal Register; 63(95):27354-27355.

U.S. Nuclear Regulatory Commission (USNRC), 1999. Standard Review Plan for Transportation Packages for Radioactive Materials. NUREG-1609. Washington, D.C.: Nuclear Regulatory Commission.

Vaughn, P., P. Swift, M. Lord, and J. Bean, 1998. Technical Review Comment Resolution of Sensitivity of the Length of Fracture Approximation in BRADFLO to Grid Used in Supplementary Analysis of the Effect of Salt Water Disposal and Waterflooding of the WIPP. Memorandum to Margaret S. Y. Chu, February, 1998. Albuquerque, N.M.: Sandia National Laboratories.

Vaughn, P., J. E. Bean, J. C. Helton, M. E. Lord, R. J. MacKinnon, and J. D. Schreiber, 2000. Uncertainty and Sensitivity Analysis for the Two-Phase Flow in the Vicinity of the Repository in the 1996 Performance Assessment for the Waste Isolation Pilot Plant: Undisturbed Conditions. Reliability Engineering and System Safety. Special Issue: The 1996 Performance Assessment for the Waste Isolation Pilot Plant; 69(1-3):205-226.

Westinghouse News, 2001a. WIPP Instructors to Provide Specialized Training for Indiana Emergency Response Professionals, January 25, 2001 [on-line]. Available: http://www.wipp.carlsbad.nm.us/pr/2001/INSTEP.pdf [February 12, 2001].

Westinghouse News, 2001b. WIPP Instructors to Provide Specialized Training for Colorado Emergency Response Professionals, January 19, 2001 [on-line]. Available: http://www.wipp.carlsbad.nm.us/pr/2001/COSTEP.pdf [February 12, 2001].

Westinghouse News, 2001c. WIPP Instructors to Provide Specialized Training for Louisiana Emergency Response Professionals, January 15, 2001 [on-line]. Available: http://www.wipp.carlsbad.nm.us/pr/2001/LASTEP.pdf [February 12, 2001].

Westinghouse News, 2001d. WIPP to Provide Training for Hospitals Personnel in Nevada, January 15, 2001 [on-line]. Available: http://www.wipp.carlsbad.nm.us/pr/2001/NVTraining.pdf [February 12, 2001].

Zhang, P. C., H. L. Anderson, Jr., J. W. Kelly, J. L. Krumhansl, and H. W. Papenguth, 1999. Kinetics and Mechanisms of Formation of Magnesite from Hydromagnesite in Brine. July 28, 1999. SAND 87185-0733. Albuquerque, N.M.: Sandia National Laboratories.

APPENDIXES

Appendix A1

Interim Report

IMPROVING OPERATIONS AND LONG-TERM SAFETY OF THE
WASTE ISOLATION PILOT PLANT

INTERIM REPORT

Committee on the Waste Isolation Pilot Plant

Board on Radioactive Waste Management

Commission on Geosciences, Environment, and Resources

National Research Council

NATIONAL ACADEMY PRESS
Washington, D.C.

NOTICE: The project that is the subject of this interim report was approved by the Governing Board of the National Research Council, whose members are drawn from the councils of the National Academy of Sciences, the National Academy of Engineering, and the Institute of Medicine. The members of the committee responsible for the report were chosen for their special competences and with regard for appropriate balance.

Support for this study was provided by the U.S. Department of Energy, under Grant No. DE-FC01-94EW54069. All opinions, findings, conclusions, and recommendations expressed herein are those of the authors and do not necessarily reflect the views of the Department of Energy.

International Standard Book Number: 0-309-06928-9

Additional copies of this report are available from:
National Academy Press
2101 Constitution Avenue, N.W.
Box 285
Washington, DC 20055
800-624-6242
202-334-3313 (in the Washington Metropolitan Area)
http://www.nap.edu

Copyright 2000 by the National Academy of Sciences. All rights reserved.

Printed in the United States of America.

THE NATIONAL ACADEMIES

National Academy of Sciences
National Academy of Engineering
Institute of Medicine
National Research Council

The **National Academy of Sciences** is a private, nonprofit, self-perpetuating society of distinguished scholars engaged in scientific and engineering research, dedicated to the furtherance of science and technology and to their use for the general welfare. Upon the authority of the charter granted to it by the Congress in 1863, the Academy has a mandate that requires it to advise the federal government on scientific and technical matters. Dr. Bruce M. Alberts is president of the National Academy of Sciences.

The **National Academy of Engineering** was established in 1964, under the charter of the National Academy of Sciences, as a parallel organization of outstanding engineers. It is autonomous in its administration and in the selection of its members, sharing with the National Academy of Sciences the responsibility for advising the federal government. The National Academy of Engineering also sponsors engineering programs aimed at meeting national needs, encourages education and research, and recognizes the superior achievements of engineers. Dr. William A. Wulf is president of the National Academy of Engineering.

The **Institute of Medicine** was established in 1970 by the National Academy of Sciences to secure the services of eminent members of appropriate professions in the examination of policy matters pertaining to the health of the public. The Institute acts under the responsibility given to the National Academy of Sciences by its congressional charter to be an adviser to the federal government and, upon its own initiative, to identify issues of medical care, research, and education. Dr. Kenneth I. Shine is president of the Institute of Medicine.

The **National Research Council** was organized by the National Academy of Sciences in 1916 to associate the broad community of science and technology with the Academy's purposes of furthering knowledge and advising the federal government. Functioning in accordance with general policies determined by the Academy, the Council has become the principal operating agency of both the National Academy of Sciences and the National Academy of Engineering in providing services to the government, the public, and the scientific and engineering communities. The Council is administered jointly by both Academies and the Institute of Medicine. Dr. Bruce M. Alberts and Dr. William A. Wulf are chairman and vice chairman, respectively, of the National Research Council.

Committee on the Waste Isolation Pilot Plant

B. JOHN GARRICK, *Chair*, PLG, Incorporated (retired), Laguna Beach, California
MARK D. ABKOWITZ, Vanderbilt University, Nashville, Tennessee
ALFRED W. GRELLA, Grella Consulting, Locust Grove, Virginia
MIKE P. HARDY, Agapito Associates, Inc., Grand Junction, Colorado
STANLEY KAPLAN, Bayesian Systems Inc., Rockville, Maryland
HOWARD M. KINGSTON, Duquesne University, Pittsburgh, Pennsylvania
W. JOHN LEE, Texas A&M University, College Station
MILTON LEVENSON, Bechtel International, Inc. (retired), Menlo Park, California
WERNER F. LUTZE, University of New Mexico, Albuquerque
KIMBERLY OGDEN, University of Arizona, Tucson
MARTHA R. SCOTT, Texas A&M University, College Station
JOHN M. SHARP, JR., The University of Texas, Austin
PAUL G. SHEWMON, Ohio State University (retired), Columbus
JAMES WATSON, JR., University of North Carolina, Chapel Hill
CHING H. YEW, The University of Texas (retired), Austin

Board on Radioactive Waste Management Liaison

DARLEANE C. HOFFMAN, Lawrence Berkeley National Laboratory, Oakland, California

Staff

KEVIN D. CROWLEY, Director
THOMAS E. KIESS, Study Director
ANGELA R. TAYLOR, Senior Project Assistant

Board on Radioactive Waste Management

JOHN F. AHEARNE, Chair, Sigma Xi and Duke University, Research Triangle Park, North Carolina
CHARLES MCCOMBIE, *Vice-Chair*, Consultant, Gipf-Oberfrick, Switzerland
ROBERT M. BERNERO, Consultant, Bethesda, Maryland
ROBERT J. BUDNITZ, Future Resources Associates, Inc., Berkeley, California
GREGORY R. CHOPPIN, Florida State University, Tallahassee
JAMES H. JOHNSON, JR., Howard University, Washington, D.C.
ROGER E. KASPERSON, Clark University, Worcester, Massachusetts
JAMES O. LECKIE, Stanford University, Stanford, California
JANE C.S. LONG, Mackay School of Mines, University of Nevada, Reno
ALEXANDER MACLACHLAN, E.I. du Pont de Nemours & Company (retired), Wilmington, DE
WILLIAM A. MILLS, Oak Ridge Associated Universities (retired), Olney, Maryland
MARTIN J. STEINDLER, Argonne National Laboratories (retired), Argonne, Illinois
ATSUYUKI SUZUKI, University of Tokyo, Japan
JOHN J. TAYLOR, Electric Power Research Institute (retired), Palo Alto, California
VICTORIA J. TSCHINKEL, Landers and Parsons, Tallahassee, Florida
MARY LOU ZOBACK, U.S. Geological Survey, Menlo Park, California

Staff

KEVIN D. CROWLEY, Director
ROBERT S. ANDREWS, Senior Staff Officer
THOMAS E. KIESS, Senior Staff Officer
GREGORY H. SYMMES, Senior Staff Officer
JOHN R. WILEY, Senior Staff Officer
SUSAN B. MOCKLER, Research Associate
TONI GREENLEAF, Administrative Associate
LATRICIA C. BAILEY, Senior Project Assistant
MATTHEW BAXTER-PARROTT, Project Assistant
LAURA D. LLANOS, Senior Project Assistant
ANGELA R. TAYLOR, Senior Project Assistant

Commission on Geosciences, Environment, and Resources

GEORGE M. HORNBERGER (Chair), University of Virginia, Charlottesville
RICHARD A. CONWAY, Union Carbide Corporation (Retired), S. Charleston, West Virginia
LYNN GOLDMAN, Johns Hopkins School of Hygiene and Public Health, Baltimore, Maryland
THOMAS E. GRAEDEL, Yale University, New Haven, Connecticut
THOMAS J. GRAFF, Environmental Defense, Oakland, California
EUGENIA KALNAY, University of Maryland, College Park
DEBRA KNOPMAN, Progressive Policy Institute, Washington, DC
BRAD MOONEY, J. Brad Mooney Associates, Ltd., Arlington, Virginia
HUGH C. MORRIS, El Dorado Gold Corporation, Vancouver, British Columbia
H. RONALD PULLIAM, University of Georgia, Athens
MILTON RUSSELL, Joint Institute for Energy and Environment and University of Tennessee (Emeritus), Knoxville
ROBERT J. SERAFIN, National Center for Atmospheric Research, Boulder, Colorado
ANDREW R. SOLOW, Woods Hole Oceanographic Institution, Woods Hole, Massachusetts
E-AN ZEN, University of Maryland, College Park
MARY LOU ZOBACK, U.S. Geological Survey, Menlo Park, California

Staff

ROBERT M. HAMILTON, Executive Director
GREGORY H. SYMMES, Associate Executive Director
JEANETTE SPOON, Administrative and Financial Officer
DAVID FEARY, Scientific Reports Officer
SANDI FITZPATRICK, Administrative Associate
MARQUITA SMITH, Administrative Assistant/Technology Analyst

Acknowledgments

This report has been reviewed in draft form by individuals chosen for their diverse perspectives and technical expertise, in accordance with procedures approved by the National Research Council (NRC) Report Review Committee. The purpose of this independent review is to provide candid and critical comments that will assist the institution in making the published report as sound as possible and to ensure that the report meets institutional standards for objectivity, evidence, and responsiveness to the study charge. The review comments and draft manuscript remain confidential to protect the integrity of the deliberative process. We wish to thank the following individuals for their participation in the review of this report:

Tom Borak, Colorado State University
Edith Boyden, Volpe National Transportation Systems Center
Robert Budnitz, Future Resources Associates, Inc.
Allen Glazner, University of North Carolina at Chapel Hill
Lawrence Johnson, National Cooperative for the Disposal of Radioactive Waste
Joseph Leary, Independent Consultant
Solomon Levy, Levy & Associates
Hank Mevzelaar, University of Utah
Randall Seright, New Mexico Institute of Technology

Although the individuals listed above have provided constructive comments and suggestions, they were not asked to endorse the conclusions or recommendations, nor did they see the final draft of the report before its release. The review of this report was overseen by E-an Zen, appointed by the Commission on Geosciences, Environment, and Resources, and Frank Parker, appointed by the Report Review Committee, who were responsible for making certain that an independent examination of this report was carried out in accordance with NRC procedures and that all review comments were carefully considered. Responsibility for the final content of this report rests entirely with the authoring committee and the NRC.

Preface

This report is the product of a National Research Council (NRC) committee study sponsored by the U. S. Department of Energy (DOE). The first NRC Committee on the Waste Isolation Pilot Plant (WIPP) began in 1978, and this committee and its successors issued eight letter reports during 1979-1992 and two full reports in 1984 and 1996. The current WIPP committee study is operating under a revised statement of task (see box) derived from a DOE request (Dials, 1997). This interim report addresses selected issues associated with the task statement, as explained below. The committee will comprehensively address the statement of task in the final report.

The specific approach taken in this interim report was to consider how to assess (1) the performance of WIPP in isolating waste from the environment and (2) the basic, minimal requirements and procedures that should be applied to waste management operations. The committee provides recommendations on several issues that it believes merit immediate consideration and action by DOE. Specifically, these issues include the determination of the natural background radioactivity in the area surrounding WIPP, and improvements in TRU waste operations.

This study is organized within the NRC's Board on Radioactive Waste Management and is being conducted by a 15-member committee. Committee members were chosen for their expertise in relevant technical disciplines such as nuclear engineering, health physics, chemical and environmental engineering, civil and transportation engineering, performance assessment, analytical chemistry, materials science and engineering, plutonium geochemistry, hydrogeology, rock and fracture mechanics, petroleum engineering, and mining engineering. As is normal practice of the National Academies, committee members do not represent the views of their institutions, but form an independent body to author this report.

To conduct the study and prepare this interim report, the committee gathered information principally through meetings and reviews of relevant literature. The committee met several times in open public sessions to hear from DOE and its contractors, as well as from other invited speakers such as regulatory agency personnel and groups with an interest in the WIPP program. Committee members prepared this report using these inputs together with their collective knowledge and experience. The report reflects a consensus of the committee and has been reviewed in accordance with NRC procedures.

> **Statement of Task**
>
> The purpose of this study is to identify the limiting technical components of the WIPP program, with a two-fold goal of (i) improving the understanding of long-term performance of the repository and (ii) identifying technical options for improvements to the National TRU Program (i.e., the engineering system that defines TRU waste handling operations that are needed for these wastes to go from their current storage locations to the final repository destination) without compromising safety.
>
> To accomplish this goal, the study will address two major issues:
>
> (1) The first is to identify research activities that would enhance the assessment of long-term repository performance. This study would examine the performance assessment models used to calculate hypothetical long-term releases of radioactivity, and would suggest future scientific and technical work that could reduce uncertainties.
>
> (2) The second is to identify areas for improvement in the TRU waste management system that may increase system throughput, efficiency, cost effectiveness, or safety to workers and the public. This study will examine, among other inputs, the current plans for TRU waste handling, characterization, treatment, packaging, and transportation.

Contents

Summary, 1
Introduction, 5
Baseline Radiogenic Analysis of Subsurface Fluids, 7
Transuranic Waste Management Program, 13
References, 27
Appendixes
    A. Background Information, 31
    B. Joint USNRC and EPA Guidance on Mixed Waste, 37
    C. Biographical Sketches of Committee Members, 39
    D. Acronyms, 44

Summary

The National Research Council convened a committee of experts to advise the U.S. Department of Energy (DOE) on the operation of the Waste Isolation Pilot Plant (WIPP), a geologic repository for disposal of defense transuranic (TRU) waste near Carlsbad, New Mexico. The committee was asked to provide recommendations on the following two issues: (1) a research agenda to enhance confidence in the long-term performance of WIPP; and (2) increasing the throughput, efficiency, and cost-benefit without compromising safety of the National TRU Program for characterizing, certifying, packaging, and shipping waste to WIPP.

The committee has written this interim report to provide DOE with recommendations on several issues that the committee believes merit immediate consideration and action. In developing this report, the committee has been guided by the principle of "reasonableness" with respect to risks, costs, and the ALARA (as low as reasonably achievable) principle. In the committee's judgment, implementing the recommendations contained in this report will contribute to the continued safe operation of WIPP. The committee will provide a more comprehensive response to its task statement (see the Preface) in the final report, which is scheduled for completion in the spring of 2001.

Research to Enhance Confidence in Long-Term Repository Performance

There has been extensive monitoring of radioactivity in the air, soils, fluvial sediments, surface water, and shallow groundwater in the area surrounding WIPP. However, the committee has determined that radiological baseline information is not available for subsurface brines and hydrocarbons near the WIPP site. This baseline information is important for environmental monitoring in the operational and post-operational phases of the repository.

Recommendation: The committee recommends that DOE should develop and implement a plan to sample oil-field brines, petroleum, and solids associated with current hydrocarbon production to assess the magnitude and variability of naturally occurring radioactive material

(NORM) in the vicinity of the WIPP site. Samples should be collected and analyzed for the radionuclides that will be present in transuranic waste emplaced at WIPP and the radionuclides common in NORM. These samples should be archived to permit subsequent analysis for constituents that may be of interest in the future. The committee recommends that a sampling plan be implemented prior to the closure of any underground rooms in WIPP that contain TRU waste.

Improvements to the National TRU Program

The National TRU Program is administered by the DOE Carlsbad Area Office and is designed to meet all applicable external regulations and internal requirements associated with the characterization, certification, packaging, and transportation of waste to WIPP. A reasonable goal for the National TRU Program is to send DOE TRU waste to WIPP at a minimum risk (from all sources of risk, including radiological exposure and highway accidents) and cost. The current system for managing TRU wastes does not achieve this goal. **The committee recommends that waste management procedures be reviewed and revised, with reduction of risk and cost as the guiding principles.**

The committee offers recommendations in this interim report to improve the following three aspects of the National TRU Program: (1) waste characterization and packaging requirements, (2) gas generation, and (3) the transportation system.

Waste Characterization and Packaging Requirements

The committee found inadequate legal or safety bases for some of the National TRU Program requirements and specifications. That is, some waste characterization specifications have no basis in law, the safe conduct of operations to emplace waste in WIPP, or long-term performance requirements. The National TRU Program waste characterization procedures involve significant resources (e.g., expenditures of several billion dollars) and potential for exposure of workers to radiation and other hazards. Insofar as some of this waste characterization may be unnecessary, such characterization is inconsistent with economic efficiency or the ALARA principle that guides radiation protection practices.

Recommendation: DOE should eliminate self-imposed waste characterization requirements that lack a legal or safety basis. One way to justify a reduction in waste characterization requirements is through implementation of joint U.S. Nuclear Regulatory Commission–U.S. Environmental Protection Agency guidance (62 Federal Register 62079; see Appendix B), which appears to the committee to provide appropriate guidelines for implementation and integration of Resource Conservation and Recovery Act (RCRA) requirements for mixed TRU waste. Another way to justify a reduction is to identify the origins of all waste characterization requirements and to eliminate those requirements that lack a technical or safety basis. Such reductions may require modifications to existing permits granted by external regulating authorities such as the Environmental Protection Agency and New Mexico Environment Department.

Gas Generation

The extreme assumptions used in DOE's current gas generation model results in gross overestimates of hydrogen (H_2) concentrations in waste packages to be shipped to WIPP. As a consequence, DOE plans to repackage some of the waste to dilute the hydrogen-producing components. These repackaging operations result in additional risks of radiation exposure to workers and highway accidents, the latter due to the increased number of truckload shipments required to transport waste in diluted form.

Recommendations:

1. DOE should derive a more realistic radiolytic gas generation model, validate it through confirmatory testing, use the results to recalculate gas generation limits, and seek regulatory approval to implement them.

2. DOE should perform a safety analysis to determine the concentration and quantity of hydrogen that, upon ignition, could damage the seals of the TRUPACT-II shipping container. The goal of the safety analysis would be to demonstrate whether such an event could occur inside a waste package, and whether the energy associated with such an event could result in rupturing the containment provided by the TRUPACT-II. This analysis could provide the rationale to obtain relief from the 5 percent hydrogen flammability limit and should form the basis for a future modification to the present TRUPACT-II license.

3. DOE should consider technical approaches for reducing hazards from hydrogen generation, such as filling the headspace of the waste containers or the shipping containers with an inert gas.

4. DOE should reevaluate the technical and regulatory feasibility of shipping high-wattage TRU waste using a railcar shipping system.

The goal of these recommendations is to expedite the transport of TRU waste to WIPP by increasing the amount of waste that can be safely carried in each truckload or trainload, without compromising the level of safety and containment that is provided by the shipping container. These recommended options would reduce the number of truckloads required to transport the waste to WIPP and the associated transportation risks.

Transportation Communication and Notification

DOE bases its system of communication and notification on the TRANSportation tracking and COMmunication (TRANSCOM) system, a satellite-based system initially developed more than a decade ago and used to track all DOE shipments of radioactive materials. Users have found the current level of performance of TRANSCOM to be less than fully reliable. Although efforts are being made by DOE to keep the system current, it has not kept pace with the rapid development of information technology. As a result, the TRANSCOM system is obsolete when compared to presently available communications systems.

Recommendations: DOE should consider cost-effective ways to improve the reliability and ease of use of the TRANSCOM system, either by improving or replacing it. If DOE decides to replace the current system, the committee strongly encourages the use or adaptation of existing commercial systems. In the near term, the DOE should develop an interim plan for maintaining an adequate communication and notification system until any such alternative system or TRANSCOM upgrade is ready for full-scale implementation. This plan should be driven by a comprehensive assessment of TRANSCOM component performance based on anticipated usage. In the long term, DOE should ensure that the system it employs is designed to meet the needs of WIPP shipment users and other major stakeholders in a timely and cost-effective fashion.

Transportation Emergency Response

The responsibility for emergency response is divided between DOE and the states along WIPP shipment corridors. In the committee's view, a system to maintain up-to-date information on response capability would contribute significantly to the effectiveness of the transportation system. The WIPP emergency response program has not assessed sufficiently whether adequate and timely emergency response coverage for a transportation incident exists along the full extent of each WIPP route. No formal system presently exists to identify areas where coverage may be inadequate.

Recommendations: The committee recommends that DOE explore with states and other interested parties how to develop processes and tools for maintaining up-to-date spatial information on the location, capabilities, and contact information of responders, medical facilities, recovery equipment, regional response teams, and other resources that might be needed to respond to a WIPP transportation incident. This assessment should explore which organization(s) should develop and maintain the capability to generate and maintain such information. DOE should also determine where emergency response capability is currently lacking, identify organization(s) responsible for addressing these deficiencies, and take action to address them.

Introduction

The Waste Isolation Pilot Plant (WIPP)[1] is a series of excavations in a Permian-age bedded salt formation approximately 660 m below the surface near Carlsbad, New Mexico (see Figure 1). Since the mid-1970s, this site has been studied for use as a geologic repository for the disposal of transuranic[2] (TRU) waste resulting from the nation's defense program. This waste contains transuranic isotopes, predominantly plutonium isotopes, which are characteristically long-lived radionuclides and therefore a long-term safety hazard. Removing these wastes from the biosphere, for example, through isolation in geologic repositories, is an appropriate strategy for protection of human health and the environment.

At WIPP, packaged waste is disposed by emplacing it in rooms excavated in the salt. Because salt under pressure flows (or "creeps") and because of the underground pressure exerted on the room ceiling, floor, and walls, over time the salt rock at these surfaces will consolidate around the waste. In time, the salt heals so as to be essentially impermeable, isolating the waste-filled rooms from the rest of the environment.

WIPP is the first deep geological repository that has been designed and engineered for radioactive waste disposal and approved by an external regulatory authority. Operations at WIPP to receive TRU waste and emplace it underground began in 1999, when TRU waste shipments were received from three U.S. Department of Energy (DOE) sites. Drums of TRU waste from the Los Alamos National Laboratory, the Idaho National Engineering and Environmental Laboratory, and the Rocky Flats Environmental Technology Site were first sent to WIPP in March, April, and June 1999, respectively.

The committee has prepared this report to provide findings and recommendations that it considers important for the safe and cost-effective operation of WIPP. The perspective of the committee has been the establishment of "reasonableness" with respect to risks, costs, and the ALARA (as low as reasonably achievable) principle (see footnote 8). The committee believes that the implementation of these recommendations will contribute to the continued safe operation of WIPP.

---

[1] A complete list of acronyms used in this report appears in Appendix D.
[2] Transuranic waste contains radionuclides with atomic numbers greater than 92 and half-lives greater than 20 years in concentrations exceeding 100 nanocuries per gram.

As noted in the preface to this report, the first component of the statement of task is "to identify research activities that would enhance the assessment of long-term repository performance" (see Appendix A). The committee considers that data from radiological site characterization measurements would provide a necessary baseline to compare against future measurements, should the integrity of WIPP ever be challenged. This issue is explored in the next section.

The second component of the statement of task pertains to improvements of the DOE TRU waste management system. To address this issue, the committee sought to identify the technical, regulatory, legal, and/or safety bases of waste management activities that significantly impacted the overall system throughput, efficiency, cost, and safety. These issues are addressed in the last section of this report.

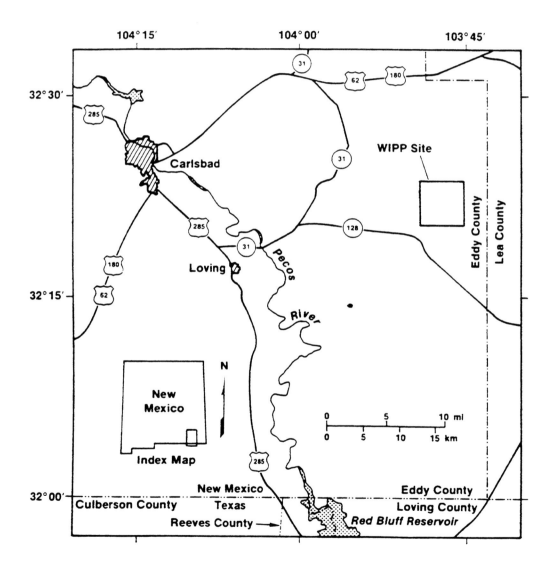

FIGURE 1 Location of the Waste Isolation Pilot Plant. Inset shows the approximate location of the map area in New Mexico. SOURCE: NRC (1996, Figure 1.1.).

Baseline Radiogenic Analysis of Subsurface Fluids

In this section the committee provides recommendations on research activities to enhance confidence in the long-term performance of WIPP. In particular, the committee considered how "baseline" studies undertaken during the early phases of repository operation could be used to support future efforts to assess repository performance.

Finding: There has been extensive monitoring of radioactivity in the air, soils, fluvial sediments, surface water, and shallow groundwater in the area surrounding WIPP.[3] However, the committee has determined that radiological baseline information is not available for subsurface brines and hydrocarbons near the WIPP site. This baseline information is important for environmental monitoring in the operational and post-operational phases of WIPP.

Recommendation: The committee recommends that DOE should develop and implement a plan to sample oil-field brines, petroleum, and solids associated with current hydrocarbon production to assess the magnitude and variability of naturally occurring radioactive material (NORM) in the vicinity of the WIPP site. Samples should be collected and analyzed for the radionuclides that will be present in transuranic waste emplaced at WIPP and the radionuclides common in NORM. These samples should be archived to permit subsequent analysis for constituents that may be of interest in the future.[4,5] The committee recommends that a

---

[3] See, for example, Conley (1999); DOE (1997c); Herczeg et al. (1988); and Kenney et al. (1999). Additionally, previous Environmental Evaluation Group studies on radiation monitoring of air, surface soil, and biota samples near the WIPP site include Neill et al. (1998); Kenney et al. (1990, 1995, 1998); Kenney and Ballard (1990); and Kenney (1991, 1992, 1994). Ramey (1985) summarizes U.S. Geological Survey data on simple radiological characterization (i.e., gross alpha, gross beta, dissolved radium, and dissolved uranium) of fluids in the Rustler Formation. References to other studies are contained in annual reports of the Carlsbad Environmental Monitoring Research Center (CEMRC, 1999) and on the CEMRC website, http://www.cemrc.org.

[4] A reanalysis of a sample using a different detection method could yield a different value. These detection limitations should be understood and distinguished from true natural differences in background radiation.

sampling plan be implemented prior to the closure of any underground rooms in WIPP that contain TRU waste.

Rationale: Early studies discounted the potential for hydrocarbon production in the vicinity of WIPP, but over the past 20 years this way of thinking has changed dramatically. The site is now surrounded by wells (see Figure 2) for hydrocarbon production (Broadhead et al., 1995), and drilling activities continue. Furthermore, it is relatively common for brines associated with hydrocarbons to be radiogenic (Bloch and Key, 1981; Fisher, 1995). Oil-field brines in the Delaware Basin share this property (Fisher, 1995). The information available on oil-field brines and petroleum resources generally consists of gross radiation measurements (i.e., gross activity), rather than analytical data on the radionuclide constituents. Such analytical data on the radioactivity of oil-field brines and petroleum resources at the WIPP site have not been made available to the committee and may not exist.

If, during or after WIPP operations, increased radioactivity in the vicinity of WIPP is observed, is this the result of a failure of the WIPP to contain its waste, or is it due to NORM? This question cannot be answered easily unless the oil-field brines, petroleum, and solids associated with hydrocarbon production (e.g., suspended solids, precipitated scale, sludges, and formation fragments) are analyzed for their naturally occurring radiation. Analyses for radioactivity and radionuclides will be necessary if disputes arise about potential releases of radionuclides from the repository. An example of the need to obtain adequate NORM background data already has been observed with occurrences of natural surface contamination on the exterior of truck transportation packages while en route to WIPP during the first three months of operation.

"Human intrusion scenarios" involving hydrocarbon exploration and production are now considered processes through which radionuclides might be released from WIPP (Kirkes, 1998). If brines have a measurable NORM content, then human intrusion that results in brine flow through WIPP to the surface is a means by which radioactivity could be carried to the surface that is not due to the TRU waste emplaced in WIPP. If oil-field brine NORM is present, then it is conceivable that NORM releases would be greater than releases from the TRU waste contents of WIPP, even if drilling breaches the repository.

Transport and disposal of oil-field brines that have high NORM contents are also potential mechanisms for localized increases in radiation. Any such increases in radiation in the vicinity of WIPP cannot necessarily be attributed to WIPP operations or the failure of WIPP to contain its waste.

There are data suggesting that oil-field brines near WIPP might contain NORM. Otto (1989, reproduced in Fisher, 1995; see Figure 3)

---

[5] The archiving of monitoring data, as well as samples, is also a long-term challenge due to the evolution of information technology and the changes in state-of-the-art storage media that will likely take place over the three decades in which WIPP is projected to be open and operational. Any data records not in paper form would be subject to such challenges.

APPENDIX A1: INTERIM REPORT

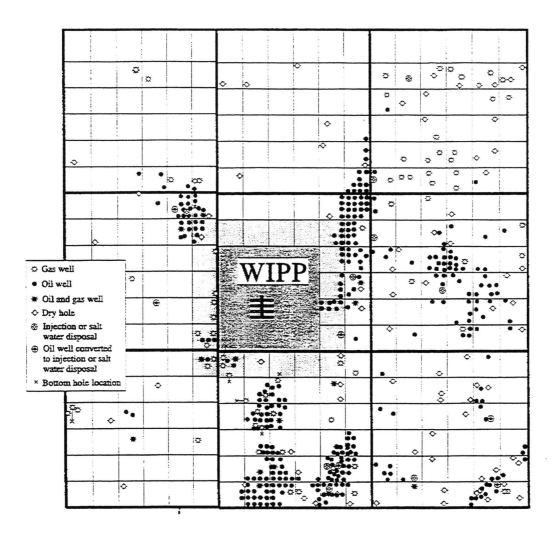

FIGURE 2 Petroleum wells in the vicinity of the WIPP site. See Figure 1 for an inset map showing the WIPP site's approximate location within New Mexico. SOURCE: Silva (1996, p. 24).

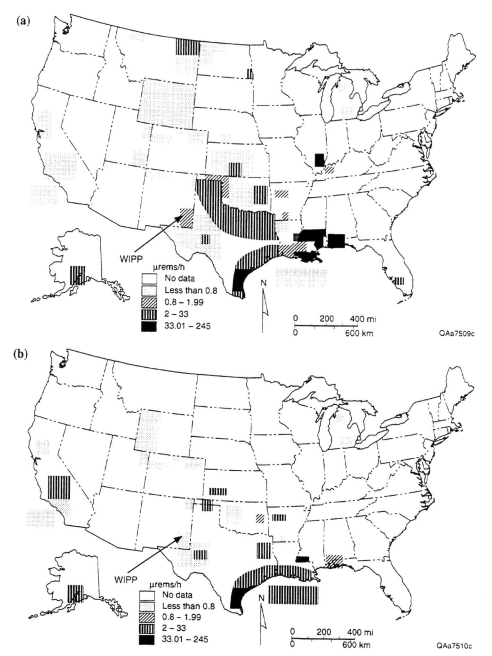

FIGURE 3 Regions of high activity from NORM in the United States from (a) oil-producing facilities and (b) gas-producing facilities. Values are aggregated median differences over background. The legend shows various shadings corresponding to various ranges of dose rates measured in microrems per hour (μrem/h). These dose rates are radioactivity measurements of NORM deposits in piping and in fluids brought to the surface. These measurements describe the concentration of radioactive species, a characteristic of the NORM deposits at any locality that is not directly dependent on the local production rate (of hydrocarbons or brine) or on the amounts of fluid that were extracted to produce the deposits. SOURCE: Fisher (1995), after Otto (1989).

shows the Delaware Basin of southeast New Mexico as a region of NORM activity in oil-and gas-producing facilities. Fisher states that "(1) not every major oil or gas field has associated high NORM levels, and (2) no major hydrocarbon-producing basin in Texas is exempt from high levels of radioactivity." The major hydrocarbon-producing basins in Texas described by Fisher include the Delaware Basin, which contain producing formations near the WIPP site, and the adjacent Central Basin Platform (Hill, 1996, p. 26).

In response to committee requests for information, DOE has answered that no data have been collected on "naturally occurring radionuclides in the underground brines and hydrocarbons near WIPP by DOE. In addition, DOE is unaware of any related information collected by the oil and gas industry" (Mewhinney, 1998b).

The need for these data is clear—no effective monitoring of the WIPP area can be successful without understanding potential sources of radiation in the environment. Air, soils, sediments, ground and surface waters, biota, and people have been analyzed to provide a database (e.g., through CEMRC activities). NORM from local hydrocarbon operations must also be analyzed. The NORM data will

- identify sources of future contamination events that might (wrongly) be attributed to a failure of WIPP;
- place any radioactivity releases from human intrusion scenarios (e.g., from petroleum exploration and production) in perspective; and
- improve the monitoring efforts.

The committee recommends near-term action to collect and analyze these data based on an appropriate sampling plan. The plan must include frequency of sampling and analyses; radionuclides to be analyzed; collection of data to assess NORM radioactivity and to estimate its variability; sampling, analysis, and archiving protocols; and producing formations to be tested. These formations should include both past (if applicable) and present producing zones, new producing zones as they become exploited in the future, and formations from which brine is (or likely will be) extracted.

Samples could come from ongoing well-based operations that generate separator streams of oil, gas, and water. These separators and separator streams are owned by the operators of the leases. The drilling of new wells would be justified if data from separator streams prove to be inadequate.

The radionuclides of interest include both those that contribute to the site's NORM background radioactivity and those in the DOE TRU inventory destined for WIPP. The NORM activity may include contributions from potassium-40, isotopes of uranium and thorium, and daughter products such as isotopes of radium. Radionuclides in TRU waste include isotopes of uranium and TRU elements and, in remote-handled[6] TRU

---

[6]Remote-handled waste is classified as that with a surface dose rate greater than or equal to 200 mrem per hour. Such waste contains fission products and activation products such as cobalt-60, strontium-90, yttrium-90, ruthenium-106,

waste, fission and activation products. Since some TRU inventory radionuclides are not commonly found in nature, sampling to determine whether such radionuclides are present in the environment may be a good way to distinguish radioactivity due to NORM from that due to TRU waste.

For the reasons given above, the committee supports the collection of NORM data on deep subsurface fluids, even though the isotopic signatures of NORM and TRU waste radioactivity are expected to differ and therefore to be readily distinguishable. In the committee's view, DOE would be better served to possess these NORM data prior to any reported discovery of significant radioactivity in the region; hence, in its recommendation the committee proposes that this survey to sample deep subsurface fluids be conducted in the near term. This survey need not continue once the measurement objectives, as proposed in this recommendation, have been met.

---

cesium-137, barium-137, and europium-152. These and other radioisotopes emit penetrating beta and gamma radiation that requires shielding.

Transuranic Waste Management Program

Transuranic waste management operations are performed under the auspices of the DOE National TRU Program administered by the DOE Carlsbad Area Office. This program has been designed and developed, based on initial efforts in the 1980s and subsequent modifications, to accommodate all applicable external regulations and internal requirements that are associated with the characterization, certification, packaging, and transportation of TRU waste to WIPP. These procedures, described briefly in Appendix A, were applied in 1999 for the first contact-handled TRU waste shipments to WIPP from DOE sites that have generated and stored such waste. The remote-handled TRU waste management system is still under development and is not reviewed in this report.

The committee considered three topics associated with TRU waste management: (1) waste characterization and packaging requirements, (2) gas generation, and (3) transportation. These topics are discussed in the following subsections.

Waste Characterization and Packaging Requirements

Finding: The committee found inadequate legal or safety bases for some of the National TRU Program requirements and specifications. That is, some waste characterization specifications have no basis in law, the safe conduct of operations to emplace waste in WIPP, or long-term performance requirements.[7] The National TRU Program waste characterization procedures involve significant resources (e.g., expenditures of several billion dollars) and potential for exposure of workers to radiation and other hazards. Insofar as some of this waste characterization may be unnecessary, such characterization is inconsistent with economic efficiency and the ALARA principle that guides radiation protection practices.[8] The committee regards the 30+ years of waste emplacement op-

---

[7] A recent study (DOE, 1999c) has also shown that some waste characterization procedures are not prescribed by safety or legal requirements.

[8] ALARA requires that all operations be done with the lowest possible radiation exposure consistent with other requirements of safety and basic programmatic objectives. See, for example, 10 CFR 835, which are requirements for

erations and related worker safety issues at WIPP as posing no significant needs for waste characterization information, because no use of characterization data is made in any handling, shipping, or emplacement operations.

Recommendation: DOE should eliminate self-imposed waste characterization requirements that lack a legal or safety basis. One way to justify a reduction in waste characterization requirements is through implementation of joint U.S. Nuclear Regulatory Commission (USNRC)–U.S. Environmental Protection Agency (EPA) guidance (62 Federal Register 62079; see Appendix B), which appears to the committee to provide appropriate guidelines for implementation and integration of Resource Conservation and Recovery Act (RCRA) requirements for mixed TRU waste. Implementation of this regulatory guidance could significantly reduce the testing protocols and associated radiation exposure of personnel. Another way to justify a reduction is to identify the origins of all waste characterization requirements and to eliminate those requirements that lack a technical or safety basis. Such reductions may require modifications to existing permits granted by external regulating authorities such as the EPA and New Mexico Environment Department.

Rationale: The National TRU Program has developed waste restrictions, as described in the waste acceptance criteria (DOE, 1996a, 1999d), and requirements for waste generating sites presented in the quality assurance program plan (DOE, 1998b). These criteria and plans impose many required procedures on waste-generating sites. EPA and DOE Carlsbad Area Office audits are conducted to certify (i.e., approve for shipment) TRU waste streams. Additionally, each container of waste from a certified waste stream must be characterized, and shipping sites must prepare documentation on characterization data for each waste container. At the Los Alamos National Laboratory, the time to obtain all the requisite documentation and administrative approvals was greater than the time to process a drum of waste through the characterization and packaging protocols that had been developed. At all sites, the assembly, management, and storage of waste characterization information are resource-intensive activities, and drum handling is a major source of worker exposure. Of interest to the committee is the origin of these required procedures, because they increase the cost or risk or decrease the efficacy of operations.

The committee sought to identify the connection between the National TRU Program procedures and the various regulatory, legal, and technical requirements that the procedures should be devised to meet. The committee views these requirements in a hierarchy, at the top of which are legal and safety requirements, with regulatory specifications at the next tier, procedures proposed by DOE to meet regulatory requirements at the third tier, and the DOE protocols for these procedures at the fourth tier.

---

worker protection referenced in DOE radioactive waste management practices (specifically, in DOE Order 435.1 [DOE, 1999a]).

The approach used by the committee was to focus on six primary National TRU Program procedures representative of high-level requirements that drive operational activities in waste characterization and repackaging (see Appendix A for an overview of these activities):[9]

1. determination that the TRU waste is of defense origin;
2. sampling and analysis of homogeneous waste;
3. headspace gas sampling and analysis;
4. radioassay of the plutonium content;
5. real-time radiography; and
6. visual examination.

These procedures are incorporated into the terms of the WIPP facility's RCRA "Part B" permit, which was issued in October 1999. The EPA guidelines that are specific to RCRA requirements are presented in Appendix B. However, the committee notes that the permit terms are subject to negotiation in a regulatory permitting process, based on the procedures proposed by DOE that became accepted as meeting regulatory requirements. A recent study (DOE, 1999c) has traced these and other TRU waste characterization requirements to their root origins in either (1) Carlsbad Area Office mandates, (2) regulatory certification and permit terms, (3) regulatory requirements or DOE orders, or (4) legal requirements.

A review of these six procedures revealed that one may be interpreted too strictly by DOE and three are without a technical or legal foundation:

Procedure 1: Determination that the TRU waste is of defense origin. WIPP is limited to defense-related waste as stipulated in the Land Withdrawal Act, with defense activities defined in the Nuclear Waste Policy Act of 1982. The committee notes that this definition includes the words "in whole or in part", which can be interpreted to include mixtures of defense and nondefense waste, although DOE does not appear to take advantage of this (see DOE, 1997a; Nordhaus, 1996). That is, waste such as plutonium-238 (^{238}Pu)-contaminated scrap from a facility used for both defense and nondefense missions at Los Alamos National Laboratory would appear to qualify as defense waste under the definition, without the need for waste segregation restrictions.

Procedure 2: Sampling and analysis of homogeneous waste. DOE has written, "There is no regulatory requirement to conduct homogeneous waste sampling and analysis, however, in an effort to meet the intent of 40 CFR 264.13, WIPP has imposed additional characterization requirements on the waste generators" (Nelson, 1999a, p. 2). No operational decisions are made based on these data; that is, the results of the sampling and analysis do not affect how waste is handled, so it is not clear what justifies the additional radiation exposure risk and cost of this procedure. In the committee's view, this sampling and analysis applied only to homogeneous waste is unnecessary: If acceptable knowledge documentation

---

[9]A more comprehensive list of TRU waste characterization procedures and their origin is found in DOE (1999c).

(see Appendix A) provides sufficient characterization information for heterogeneous waste, the committee can identify no technical reason why acceptable knowledge should not also be adequate for homogeneous waste.

Procedure 3: Headspace gas sampling and analysis. DOE informed the committee that "there is no regulatory requirement to conduct headspace gas sampling and analysis, however, in an effort to meet the intent of 40 CFR 264.13, WIPP has imposed additional characterization requirements on the waste generators" (Nelson, 1999a, p. 3). The headspace gas sampling and analysis was developed as a means of checking on conformance with USNRC and the U.S. Department of Transportation (DOT) requirements (see Appendix A for relevant sections of these regulations); however, these requirements can be met by other means (see the recommendations that follow on the issue of gas generation).

Procedure 6: Visual examination. Visual examination is done on a fraction of the waste containers to confirm the real-time radiography and acceptable knowledge waste characterization information (Nelson, 1999a, p. 5). However, there is no requirement for verification of real-time radiography results. An alternative way to confirm these results without operator exposure would be to use standardized test drums. The visual examination confirmation is a self-imposed procedure that yields no benefit but results in increased risk of exposure and cost.

A DOE study (1999c) also confirms that procedures 2, 3, and 6 identified above are based on terms negotiated in a permit and not on a required regulation or legal mandate. The committee sees no utility in the information that these procedures provide. Any speculative benefits of acquiring this information must be weighed against the risks and costs. The committee's judgment is that the collection of these data from superfluous procedures increases, rather than decreases, the risk and safety of the overall TRU waste operations.

These superfluous characterization and intrusive procedures also represent a conflict with the ALARA principle. The issue of how to handle conflict between regulatory requirements for waste characterization information and ALARA is beyond the scope of the committee's statement of task. At issue, however, is whether the present TRU waste management program results in significantly more worker radiation exposure than is justified to satisfy safety and nonnegotiable regulatory requirements.

Gas Generation

Finding: The extreme assumptions used in DOE's current gas generation model result in gross overestimates of hydrogen concentrations in waste packages to be shipped to WIPP. As a consequence, DOE's plans to repackage some of the waste to dilute the hydrogen-producing components. These repackaging operations result in additional risks of radiation exposure to workers and highway accidents due to the increased number of truckload shipments required to transport waste in diluted form.

Recommendations:

1. DOE should derive a more realistic radiolytic gas generation model, validate it through confirmatory testing, use the results to recalculate gas generation limits, and seek regulatory approval to implement these limits.

2. DOE should perform a safety analysis to determine the concentration and quantity of hydrogen that, upon ignition, could damage the seals of the TRUPACT-II shipping container. The goal of the safety analysis would be to demonstrate whether such an event could occur inside a waste package, and whether the energy associated with such an event could result in the rupture of containment provided by the TRUPACT-II. This analysis could provide the rationale to obtain relief from the 5 percent hydrogen flammability limit and should form the basis for a future modification to the present TRUPACT-II license.

3. DOE should consider technical approaches for reducing hazards from hydrogen generation, such as filling the headspace of the waste containers or the shipping containers with an inert gas to displace air and thereby reduce the flammability hazard.

4. DOE should reevaluate the technical and regulatory feasibility of shipping high-wattage TRU waste using ATMX[10] railcar shipping system.

The goal of these recommendations is to expedite the transport of TRU waste to WIPP by increasing the amount of waste that can be carried safely in each truckload or trainload, without compromising the level of safety and containment that is provided by the shipping container. These recommended options would reduce the number of truckloads required to transport the waste to WIPP and the associated transportation risks.

Rationale: The amount of TRU waste in each waste drum and truck shipment is limited because of the potential for radiolytic generation of hydrogen gas (H_2). Within TRU waste, radiolytic hydrogen gas generation is due primarily to the co-disposal of alpha emitters with organic materials. The DOE has developed a radiolysis model to calculate hydrogen generation rates and the hydrogen concentration in each headspace[11] inside a waste container. Limiting any H_2 concentration to 5 percent leads to a restriction, expressed as maximum allowable wattage, on alpha activity (i.e., the amount of alpha-emitting radionuclides) within each waste container (e.g., a 55-gallon drum). The value of 5 percent H_2 (as a mole

---

[10] "ATMX" is an acronym to denote the railcars used by DOE to ship nuclear weapons components and TRU waste. The "AT" stands for Atchison Topeka, the rail carrier. The "M" signifies munitions, and the "X" on a railcar signifies private ownership (in this case, by the U.S. government), rather than ownership by the railroad company. As noted elsewhere in this report, these railcars have been used to ship TRU waste for decades.

[11] In many waste containers, waste is contained in one or more plastic bags that were used for radiological protection against any inadvertent spread of radioactivity. These plastic bags provide resistance to diffusive transport of hydrogen gas, thereby providing multiple headspaces.

fraction) in air as a "flammability limit" can be used in any USNRC license application on a transportation package without the need for further safety analysis because of its conservatism. This allowable wattage is a function of the G value[12] of the solid matrix of the waste materials adjacent to each alpha emitter and the total resistance to the flow of hydrogen gas that the waste and packaging contents provide, due primarily to the layers of plastic bags in the waste.

Wattage limits based on this model determine whether or not a waste container may be transported to WIPP without repackaging. The gas generation model, and the wattage limits derived from it, specify the terms of operation that are contained in the DOE safety analysis report for the TRUPACT-II transportation package. These terms of operation are also specified in DOE's application to the USNRC for regulatory approval of the TRUPACT-II transportation package. The certificate of compliance for TRUPACT-II issued by the USNRC is subject to modifications (and in fact has been amended several times since the original certificate was issued in the late 1980s), provided that DOE can offer sufficient adequate safety assurances and comply with applicable regulations, principally the USNRC's 10 CFR 70-71 and DOT's 49 CFR 171-173.

The current model is based on worst-case scenario of H_2 generation and wattage limits. Because of this worst-case approach and the extreme assumptions used in the model, the calculations often exceed experimental observations by orders of magnitude. The explanations for these large discrepancies are only beginning to be studied (see Idaho Engineering and Environmental Laboratory, 1998; Mewhinney, 1998a). Specific examples follow.

1. A G value of 3.4 is used for the plastic bags in the safety analysis report for the TRUPACT-II (DOE, 1997b). In this analysis, no credit is taken for matrix depletion (i.e., exhaustion of the H_2 source). Therefore, DOE is seeking relief from unrealistically large G values in revisions 17-19 of the safety analysis report and certificate of compliance for the TRUPACT-II (DOE, 1999b).

2. The model assumes that all layers of plastic bags are intact and behave as a new bag (i.e., no credit is taken for changes in permeability with age).

The results of these gas generation model assumptions have severe consequences.[13] Repackaging is carried out to redistribute waste in containers (e.g., 55-gallon drums) in order to meet the wattage limits derived from the gas generation model for each container. This repackaging of waste exposes workers to radiation and increases the number of containers, thereby diluting the waste into a greater volume.

---

[12]The G value is the number of electrons (or, equivalently, the number of electron-ion pairs, with H^+ the chief ion produced in materials containing hydrogen compounds) produced in a material per 100 eV of energy that is deposited within it by irradiation.

[13]In general, the use of extreme assumptions that result in overestimating consequences is not a conservative approach, because attending to these overestimated consequences results in unnecessary actions, each of which has its own risks, thus potentially increasing the risks of the overall operations.

Transportation-related risks (and costs) are also incurred in repackaging, because the extra containers require additional shipping loads with many additional truck trips. DOE estimates reveal that this repackaging of ^{238}Pu contact-handled TRU waste may increase the number of ^{238}Pu shipments by more than a factor of ten, to as many as 150,000 extra drums (Lechel and Leigh, 1998).[14] Another consequence of such volume expansions that should be considered is the impact on WIPP's volume limit.[15] Therefore, the maximum allowable wattage imposed by the gas generation model is a major technical restriction of the National TRU Program.

Recent information (DOE, 1999b; Gregory, 1999) suggests that significant progress is being made toward developing technical information to support planned future applications to the USNRC to amend the terms of the TRUPACT-II safety analysis report and certificate of compliance. Research continues to investigate the use of hydrogen getters[16] (Mroz et al., 1997, 1999), methods for puncturing bags, use of vented bags (Gregory, 1999), and relief from the restrictive G values (Idaho Engineering and Environmental Laboratory, 1998).

To provide containment of its radioactive contents, the TRUPACT-II shipping container uses outer O-rings that generate a vacuum seal. In this package design, internally generated gas, such as H_2, builds up to pressurize the internal gas volume. Other transportation package designs are possible that are less sensitive than the TRUPACT-II to the potential for H_2 gas generation. One such system for transport of TRU wastes was the ATMX railcar system, which DOE used for hundreds of shipments over several decades to safely transport TRU waste from the Mound Laboratory in Ohio and from the Rocky Flats Environmental Technology Site in Colorado to the Idaho National Engineering and Environmental Laboratory. Based on the integrity provided by the railcar, this system was exempted (DOT exemption number DOT-E 5948) from the double-containment and vacuum seal requirements for packages used to transport plutonium (classified as "Type B" fissile packages). As a result, this system did not suffer limitations of the kind that are imposed on the TRUPACT-II due to radiolytic gas generated and trapped within the shipping container.

---

[14]The actual number of containers to be repackaged and procedures to be used have not yet been determined by DOE but are under active study, as is an analysis of technical options. If each truck carried the maximum number of TRUPACT-II transporters per shipment to WIPP, and each TRUPACT-II carried the maximum number of 55-gallon drums, 150,000 drums would be equivalent to 3,600 additional truck shipments.

[15]The Land Withdrawal Act (P.L. 102-579) specifies a total TRU waste volume limit of 175,600 m^3; if waste were sufficiently diluted, WIPP would be filled to this volume limit without having disposed the total TRU inventory in curies. Therefore, there is a minimum "filling ratio" of curies to volume that must be achieved, on average, for WIPP to contain the total TRU inventory in curies by the time the volume restriction is reached.

[16]A getter is a material designed to absorb gas such as hydrogen.

Transportation

The committee has examined various aspects of the WIPP transportation system, focusing on system safety and the cost-effectiveness of planned and ongoing activities. Based on this review (see DOE, 1999b; Mewhinney, 1998a,b), the committee has identified two issues—DOE's communication and notification system (TRANSCOM[17]) and DOE's emergency response program—that warrant immediate attention.

DOE's Communication and Notification Program

Finding: DOE bases its system of communication and notification on TRANSCOM, a satellite-based system developed more than a decade ago and used to track all DOE shipments of radioactive materials. Users have found the current level of performance of TRANSCOM to be less than fully reliable. Although efforts are being made to keep the system current (Nelson, 1999b), it has not kept pace with the rapid development of information technology. As a result, TRANSCOM is obsolete compared to presently available communications systems (for a summary of recent transportation communication initiatives using information technology, see Allen [1998]).

Recommendations: DOE should consider cost-effective ways to improve the reliability and ease of use of the TRANSCOM system, either by improving or replacing it. If DOE decides to replace the current system, the committee strongly encourages the use or adaptation of existing commercial systems. In the near term, DOE should develop an interim plan for maintaining an adequate communication and notification system until any such alternative system or TRANSCOM upgrade is ready for full-scale implementation. This plan should be driven by a comprehensive assessment of TRANSCOM component performance based on anticipated usage. In the long term, DOE should ensure that the system it employs is designed to meet the needs of WIPP shipment users and other major stakeholders in a timely and cost-effective fashion.

Rationale: Public confidence in a transportation communication and notification system is essential. This will become increasingly important with the growing number of shipments to WIPP. The magnitude of shipping activity and the public interest in WIPP transportation safety dictate the need for a state-of-the-art communications system.

As a means of obtaining information on the current effectiveness of TRANSCOM, the committee contacted 27 users located across the nation, requesting information on their experience with the system. Serious concerns were raised about system reliability and ease of use, giving the impression that key transportation stakeholders have little confidence in TRANSCOM. Comments of the 11 users who responded (from two tech-

---

[17] The DOE TRANSportation Tracking and COMmunication System, or TRANSCOM, is a satellite-based telecommunications system designed to enable users to track WIPP truck shipments in essentially real time while en route to WIPP on the approved highway routes.

nology companies and various institutions involved in emergency response monitoring in Colorado, Illinois, Pennsylvania, Idaho, Wyoming, Oregon, Arizona, North Carolina, and Utah) to a committee survey are shown below. On a scale of 1 to 5, with 1 = inadequate, 2 = poor, 3 = average, 4 = good, and 5 = excellent, the average scores for TRANSCOM system on five issues were as follows:

Category	Average Score on Scale of 1 to 5
Accuracy	3.5
Cost	3.4
Ease of use	3.2
Communication capability	3.0
Reliability	2.5

Most survey responders also wrote either explicitly or by using examples that the system was (1) unreliable (citing frequent downtime, connection or access problems, or other hardware or software problems), (2) not user friendly (citing features such as slow data rates, the time required to download information, and "old technology"), and (3) not economical because of the high costs for modem connections. Of those survey responders who had experience with at least one other transportation tracking system, each provided written comments attesting to the "unreliable" and/or "not user-friendly" features of TRANSCOM.

The committee concludes from this survey and from other materials received (e.g., presentations at committee meetings in October 1998, May 1999, and July 1999) that the TRANSCOM system has failed to give its users confidence in its reliability, ease of use, and the timeliness with which accurate information can be accessed. The committee regards these features as important for engendering public confidence and trust in WIPP's transportation program, especially for incidents in which some sort of emergency response is required.

The committee considers that given the potential interest in and visibility of WIPP shipments, the tracking system should provide reliable, real-time, and user-friendly access to information for the state users and other interested parties. In principle, this could be accomplished through upgrades to the current TRANSCOM system. However, rather than maintaining and upgrading a technically obsolete system, the committee believes that it would be more prudent for DOE to implement a less expensive, higher-quality system using a currently available commercial communications product (for a summary of transportation communication initiatives using information technology, see Allen [1998]). Careful screening of vendors is necessary to ensure that the desired system can perform to specification and be delivered on schedule and within budget.

Recent DOE efforts (Nelson, 1999b) are aimed at developing upgraded information technology capabilities ("TRANSCOM 2000") for the TRANSCOM system. Specifically, modem connections to access data of interest (e.g., the commercial bill of lading for a shipment) are to be replaced in the near future by internet postings. These plans for improved user interface and data distribution capabilities do not address other parts of the system, such as the speed with which data are acquired and processed prior to posting. These data acquisition and processing activities appear to introduce time delays that limit system performance; for example, position updates showing the locations of trucks along routes are delayed by several (up to seven) minutes (Nelson, 1999b). An as-yet-unspecified element of these planned upgrades is the extent to which future stakeholder participation will be solicited and used to provide sufficient feedback to ensure that the product ultimately developed addresses user concerns. Moreover, the timetable for off-the-shelf availability of TRANSCOM 2000 appears to the committee to be several years in the future, a problematic scenario for a WIPP shipping activity that is already underway.

One issue relevant to these planned information disclosures in TRANSCOM 2000 is the extent to which such information is needed or useful, by which parties, and to what ends. For example, the terrorist hazard and/or the potential for deliberate sabotage would presumably increase as this information is disseminated more broadly. If restricted access to certain information were important, security firewalls could be used to prevent internet information from being accessed outside of the TRANSCOM user community.

At present, the National TRU Program is one of many DOE users of the TRANSCOM system that is managed by another DOE program unit, the DOE transportation center in Albuquerque, New Mexico; other DOE transportation users include shippers of low-level waste and spent nuclear fuel. If the DOE transportation program that maintains TRANSCOM cannot provide sufficient improvements to fully implement the above recommendations, another approach would be for the National TRU Program to adapt a commercially available tracking system for use on WIPP shipments only. If the tracking system need only meet WIPP shipment requirements, the system specifications would likely be simpler, with a correspondingly greater likelihood that a commercially available product could be adapted for use. For example, WIPP shipments involve unclassified material, which may allow relief from the full suite of TRANSCOM system requirements that have been developed for all of DOE shipping needs.

DOE's Emergency Response Program

Finding: The responsibility for emergency response is divided between DOE and the states along WIPP shipment corridors. In the committee's view, a system to maintain up-to-date information on response capability would contribute significantly to the effectiveness of the transportation system. The WIPP emergency response program has not assessed sufficiently whether adequate and timely emergency response coverage for a transportation incident exists along the full extent of each WIPP route. No formal system presently exists to identify areas where coverage may be inadequate.

Recommendations: The committee recommends that DOE explore with states and other interested parties how to develop processes and tools for maintaining up-to-date spatial information on the location, capabilities, and contact information of responders, medical facilities, recovery equipment, regional response teams, and other resources that might be needed to respond to a WIPP transportation incident. This assessment should explore which organization(s) should develop and maintain the capability to generate and maintain such information. DOE should also determine where emergency response capability is currently lacking, identify organization(s) responsible for addressing these deficiencies, and take action to address them.

Rationale: To respond appropriately to any accident or other incident associated with a WIPP shipment, an emergency response system has been developed involving the DOE and state and local governments. Four levels of emergency response teams have been established. The first responders, typically the local police or local fire department, are to alert others. Their "911" call routes the incident to the attention of the second responders, the state emergency management agency, which then involves the state police and any state hazardous material (HAZMAT) or radiological response teams. The third responders are DOE Radiological Assistance Program teams that would be sent from major DOE sites (e.g., Idaho Engineering and Environmental Laboratory or the DOE Carlsbad Area Office) to conduct radiological emergency (medical) response. The fourth level of response is DOE remediation teams who perform measures such as righting a truck and any necessary site cleanup and restoration activities (DOE, 1998a).

Because of the required integrity of the TRUPACT-II shipping container, which is tested and certified to conform to the USNRC's 10 CFR 71 regulatory requirements, the containment offered by this container normally cannot be breached in an accident scenario. Therefore, emergency response procedures in these four levels of response normally would preclude any consideration of releases of materials from the TRUPACT-II. Under normal conditions, the emergency response procedures would still be needed for traffic management and other necessary operations in accident-related situations.

DOE's emergency response program relies heavily on WIPP corridor states to conduct emergency responder training and develop response plans in the event of a transportation incident. DOE also maintains its own specialized response capabilities that can be deployed on an as-needed basis. Although this approach offers certain advantages in terms of state and local involvement, system-level integration is a significant concern.

Maintaining an effective emergency response program necessitates that, if an incident should occur anywhere along a WIPP route, qualified responders can reach the scene in a timely fashion. Emergency preparedness is a formidable challenge given the thousands of miles of highway that comprise WIPP routes.

While WIPP corridor states are coordinating with DOE to ensure the safe transport of WIPP shipments[18] (DOE, 1995, 1999b; Klaus, 1999; Ross, 1999; Wentz, 1999), the public may view this responsibility as ulti-

---

[18]These activities have included training drills that have been conducted over the past several years to simulate real transportation procedures and accident scenarios.

mately resting with DOE as the system manager. The public might well expect qualified emergency response coverage along the entire length of each WIPP route, and in the committee's view, DOE could be heavily criticized if an event occurs that demonstrates weaknesses in the emergency response program, regardless of whether serious consequences are involved. Hence, although the recommendations in this section are not legal requirements, these assessments of the emergency response capabilities are, in the committee's view, important for providing a well-orchestrated transportation system.

The system-level integration necessary to ensure adequate emergency response would have to manage the jurisdictional boundaries between the various responsible government agencies. For example, under the federal Occupational Safety and Health Act (specifically, 29 CFR 1910.120), an employer is responsible for providing training; consequently, the state has the responsibility to determine the extent and adequacy of training (i.e., who is trained and in what capabilities) for first- and second-level responders. States have, to date, offered free WIPP-related training opportunities. No "quality assurance" program yet exists to evaluate periodically and systematically the extent of training and response capabilities within states. Moreover, the database lists trained personnel by state only, rather than by local region (e.g., county). As required by the Land Withdrawal Act, DOE provides the states with WIPP-specific hazard information, but DOE does not furnish protective, detection, monitoring, or communication equipment to states.

These and other demarcations of responsibilities should be managed to ensure that prompt and effective response capability for any transportation incident exists anywhere along a WIPP route. Although the training and response time associated with the first and second responders are not under DOE's direct control, a system to assess the extent and adequacy of this response coverage would be useful for DOE to properly prepare for and manage WIPP transportation incidents.

Committee Perspective on National TRU Program Requirements

A reasonable goal for the National TRU Program is to send DOE TRU waste to WIPP at a minimum risk (from all sources of risk, including radiological exposure and highway accidents) and cost. The current system for managing TRU wastes does not achieve this goal. The current transportation system cannot be used to ship a large fraction of the TRU waste volume without significant repackaging (Connolly and Kosiewicz, 1997; DOE, 1999b; Mroz et al., 1997). For the waste inventory that does qualify for shipment in this system, risk and cost considerations have not been optimized.

The terms and activities selected by DOE Carlsbad Area Office for submission to its regulatory authorities to satisfy applicable regulations and other concerns do not produce an optimum balance between risk and cost, in the spirit of ALARA. **The committee recommends that waste management procedures be reviewed and revised, with reduction of risk and cost as the guiding principles.**

As experience is gained in the WIPP shipping program, empirical data could be gathered to improve upon the initial estimates of risk and cost that are associated with each operation. The effort to reduce risks and costs necessarily would include some consideration of uncertainty, the procedures needed to adequately bound this uncertainty, and an assessment of which TRU waste program elements are the most important to control.

For example, the current National TRU Program has many procedures to control certain program elements. Over time, the most effective of such controls could be identified and retained. The reduction of risks and costs is possible in a management approach that takes into consideration public preferences for certain restrictions and implements procedures to minimize relevant uncertainties. As empirical data and experience are gathered, estimates of risks and costs of various components of the TRU waste operations can be refined. Such risk and cost estimates are useful to probe the elements of the waste management system that need to be controlled most restrictively, whether to meet legal or technical safety restrictions or to address public preferences for how radioactive waste is to be managed and transported.

References

Allen, J. C. 1998. ITS-HM incident management–system coordination. Presentation at the International Border Clearance Planning and Deployment Committee Meeting. Mexico City. October 28-29.

Bloch, S., and R. M. Key. 1981. Modes of formation of anomalously high radioactivity in oil-field brines. American Association of Petroleum Geologists Bulletin. Vol. 65: 154-159.

Broadhead, R. F., F. Luo, and S. W. Speer. 1995. Evaluation of Mineral Resources at the Waste Isolation Pilot Plant (WIPP) Site. Carlsbad, N. Mex.: Westinghouse Electric Corp. Waste Isolation Division.

Carlsbad Environmental Monitoring Research Center. 1999. 1998 Report. Waste-Management Education & Research Consortium (WERC). Carlsbad, N. Mex.: College of Engineering. New Mexico State University.

Channell, J. K., and R. Neill. 1999. A Comparison of the Risks from the Hazardous Waste and Radioactive Waste Portions of the WIPP Inventory (EEG-72 and DOE/AL58309-72). Albuquerque, New Mex.: Environmental Evaluation Group.

Conley, M. 1999. Environmental monitoring at Carlsbad Environmental Monitoring Research Center. Presentation to the Committee on the Waste Isolation Pilot Plant. Albuquerque, New Mex. July 26.

Connolly, M., and S. Kosiewicz. 1997. TRU waste transportation: The flammable gas generation problem. Technology: Journal of the Franklin Institute. Vol. 334A: 351-356.

Dials, G. 1997. Letter to Dr. Bruce Alberts. May 2.

Federal Register. Friday, June 1, 1990. Vol. 55. No. 106. P. 22669. Waste Analysis Plans and Treatment/Disposal Facility Testing Requirements.

Federal Register. Thursday, November 20, 1997. Vol. 62. No. 224. Pp. 62079-62094. Nuclear Regulatory Commission and Environmental Protection Agency Joint Guidance on Testing Requirements for Mixed Radioactive and Hazardous Waste.

Fisher, R. S. 1995. Naturally Occurring Radioactive Materials (NORM) in Produced Water and Scale from Texas Oil, Gas, and Geothermal Wells: Geographic, Geologic, and Geochemical Controls. Geological Circular. Vol. 95-3, 43 pp. Austin: University of Texas Bureau of Economic Geology.

Gregory, P. 1999. Gas generation model for TRUPACT-II. Presentation to the Committee on the Waste Isolation Pilot Plant. Albuquerque, N. Mex. July 26.

Herczeg, A. L., H. J. Simpson, R. F. Anderson, R. M. Trier, G. G. Mathieu, and B. L. Deck. 1988. Uranium and radium mobility in groundwaters and brines within the Delaware Basin, southeastern New Mexico, U.S.A. Chemical Geology. Vol. 72: 181-196.

Hill, C. A. 1996. Geology of the Delaware Basin, Guadalupe, Apache, and Glass Mountains, New Mexico and West Texas: Permian Basin Section. Midland, Tx. Society for Sedimentary Geology Publication No. 96-39, 480 pp.

Idaho National Engineering and Environmental Laboratory (INEEL). 1998. TRUPACT-II Matrix Depletion Program Final Report. INEEL/EXT-98-00987. Rev. 0. September.

Kenney, J., J. Rodgers, J. Chapman, and K. Shenk. 1990. Preoperational Radiation Surveillance of the WIPP Project by EEG, 1985-1988 (EEG-43). Albuquerque, N. Mex.: Environmental Evaluation Group.

Kenney, J. W. 1991. Preoperational Radiation Surveillance of the WIPP Project by EEG During 1990 (EEG-49). Albuquerque, N. Mex.: Environmental Evaluation Group.

Kenney, J. W. 1992. Preoperational Radiation Surveillance of the WIPP Project by EEG During 1991 (EEG-51). Albuquerque, N. Mex.: Environmental Evaluation Group.

Kenney, J. W. 1994. Preoperational Radiation Surveillance of the WIPP Project by EEG During 1992 (EEG-54). N. Mex.: Environmental Evaluation Group.

Kenney, J. W., and S. C. Ballard. 1990. Preoperational Radiation Surveillance of the WIPP Project by EEG During 1989 (EEG-47). Albuquerque, N. Mex.: Environmental Evaluation Group.

Kenney, J. W., P. S. Downes, D. H. Gray, and S. C. Ballard. 1995. Radionuclide Baseline in Soil Near Project Gnome and the Waste Isolation Pilot Plant (EEG-58). Albuquerque, N. Mex.: Environmental Evaluation Group.

Kenney, J. W., D. H. Gray, and S. C. Ballard. 1998. Preoperational Radiation Surveillance of the WIPP Project by EEG During 1993 Through 1995 (EEG-67). Albuquerque, N. Mex.: Environmental Evaluation Group.

Kenney, J. W., D. H. Gray, S. C. Ballard, and L. Chaturvedi. 1999. Preoperational Radiation Surveillance of the WIPP Project by EEG from 1996-1998 (EEG-73). Albuquerque, N. Mex.: Environmental Evaluation Group.

Kirkes, R. 1998. Resource extraction near WIPP—A status of current industry practice. Westinghouse Electric Company report to the Committee on the Waste Isolation Pilot Plant. Albuquerque, N. Mex. August 18.

Klaus, J. 1999. Presentation to the Committee on the Waste Isolation Pilot Plant. Albuquerque, N. Mex. July 26.

Lechel, D. J., and C. D. Leigh. 1998. Plutonium-238 Transuranic Waste Decision Analysis. SAND98-2629. Albuquerque, N. Mex.: Sandia National Laboratories.

Mewhinney, J. A. 1998a. Letter to Thomas Kiess. October 7, and enclosures.

Mewhinney, J. A. 1998b. Letter to Thomas Kiess. December 15, and enclosures.

Mroz, E., S. Kosiewicz, D. Finnegan, C. Leibman, S. Djordjevic, C. Loehr, and J. Weinrach. 1997. Increasing TRUPACT-II wattage limits: Two technical approaches. Technology: Journal of the Franklin Institute. Vol. 334A: 357-363.

Mroz, E., D. Finnegan, P. Noll, S. Djordjevic, C. Loehr, C. Banjac, J. Weinrach, J. Kinker, and M. Connolly. 1999. Increasing TRUPACT-II wattage limits: Hydrogen G-Values and getters. Presentation at Waste Management '99. Tucson, Ariz. March.

National Research Council. 1996. The Waste Isolation Pilot Plant: A Potential Solution for the Disposal of Transuranic Waste. Washington, D.C.: National Academy Press.

Neill, R. H., L. Chaturvedi, D. F. Rucker, M. K. Silva, B. A. Walker, J. K. Channell, and T. M. Clemo. 1998. Evaluation of the WIPP Project's Compliance with the EPA Radiation Protection Standards for the Disposal of Transuranic Waste (EEG-68). Albuquerque, N. Mex.: Environmental Evaluation Group, 291 pp., plus appendixes.

Nelson, R. 1999a. E-mail correspondence to committee containing attachment of summaries of regulatory drivers for certain characterization activities. Carlsbad, N. Mex.: Department of Energy. May 11.

Nelson, R. 1999b. E-mail correspondence to committee containing attachment of TRANSCOM and TRANSCOM 2000 Report. Carlsbad, N. Mex.: Department of Energy. October 21.

Nordhaus, R. 1996. Department of Energy memorandum. Interpretation of the term "Atomic Energy Defense Activities" as used in the Waste Isolation Pilot Plant Land Withdrawal Act. September 9.

Olson, W. C. 1999. Letter to Thomas Kiess on New Mexico NORMS regulations. May 19, with attachment.

Otto, G. H. 1989. A national survey on naturally occurring radioactive materials (NORM) in petroleum producing and gas processing facilities. Report to the American Petroleum Institute, 265 pp.

Ramey, D. S. 1985. Chemistry of Rustler Fluids (EEG-31). Environmental Evaluation Group. Albuquerque, N. Mex.: Environmental Improvement Division Health and Environment Department State of New Mexico.

Ross, R. 1999. Presentation to the Committee on the Waste Isolation Pilot Plant. Albuquerque, N. Mex. July 26.

Silva, M. K. 1996. Fluid injection for salt water disposal and enhanced oil recovery as a potential problem for the WIPP. Proceedings of a June 1995 Workshop and Analysis (EEG-62). Albuquerque, N. Mex.: Environmental Evaluation Group, 177 pp.

U.S. Department of Energy. 1995. Emergency Planning, Response, and Recovery: Roles and Responsibilities for TRU Waste Transportation Incidents. DOE/CAO-94-1039. Carlsbad, N. Mex.

U.S. Department of Energy. 1996a. Waste Acceptance Criteria for the Waste Isolation Pilot Plant. DOE/WIPP-069. Rev. 5. Carlsbad, N. Mex.

U.S. Department of Energy. 1996b. TRUPACT-II Content Codes (TRUCON). DOE/WIPP 89-004. Rev. 10. December.

U. S. Department of Energy. 1997a. Carlsbad Area Office Interim Guidance on Ensuring that Waste Qualifies for Disposal at the Waste Isolation Pilot Plant. February 13.

U.S. Department of Energy. 1997b. Safety Analysis Report for the TRUPACT-II Shipping Package. Rev. 16. February.

U.S. Department of Energy. 1997c. Waste Isolation Pilot Plant Annual Site Environmental Report Calendar Year 1996: Waste Isolation Division. Westinghouse Electric Corp. Report. DOE/WIPP 97-2225, 9 chapters.

U.S. Department of Energy. 1998a. Waste Isolation Pilot Plant Transportation Plan. DOE/CAO 98-3103. Rev. 0 November 10.

U.S. Department of Energy. 1998b. Transuranic Waste Characterization Quality Assurance Program Plan. CAO-94-1010. Rev. 1.0 December 18.

U.S. Department of Energy. 1999a. DOE Order 435.1. http://www.explorer.doe.gov:1776/htmls/reqs/doe/newserieslist.html. Washington, D.C.

U.S. Department of Energy. 1999b. DOE Responses to Requests for Information from the National Academies Committee on the Waste Isolation Pilot Plant. July 22.

U.S. Department of Energy. 1999c. Findings and Recommendations of the Transuranic Waste Characterization Task Force. Final Report. August 9.

U.S. Department of Energy. 1999d. Waste Acceptance Criteria for the Waste Isolation Pilot Plant. Revision 7. DOE/WIPP-069.

Wentz, C. 1999. Presentation to the Committee on the Waste Isolation Pilot Plant. Albuquerque, N. Mex. July 26.

Appendix A

Background Information

The material in this appendix provides background information on the long-term performance of the Waste Isolation Pilot Plant (WIPP) as well as waste characterization and transportation activities associated with the National TRU Program.

Assessment of Long-Term Performance

The ability of WIPP to isolate radioactive waste from the accessible environment has been studied and modeled in a performance assessment calculation. The performance assessment organizes information relevant to long-term (i.e., over a 10,000-year period) repository behavior by assessing the probability and consequence of major scenarios by which radionuclides can be released to the environment surrounding the WIPP site. Important scenarios include those due to human activities, whether deliberate or unintentional, that might occur near the WIPP site and potentially compromise the integrity of the repository. For example, drilling for hydrocarbon resources in formations underlying WIPP is currently practiced in the Delaware Basin on land surrounding the WIPP site; therefore, stylized "human intrusion" scenarios in which future boreholes are drilled through WIPP have been analyzed in the performance assessment model.

Using this performance assessment, the U.S. Department of Energy (DOE) has modeled the long-term performance of the WIPP repository to meet regulatory requirements. As specified by the 1992 Land Withdrawal Act (P.L. 102-579) passed by the U.S. Congress, the U.S. Environmental Protection Agency (EPA) is the external regulatory authority for WIPP, using as a regulatory standard the rule 40 CFR 191.[1] The performance assessment model formed the basis of the 1996 DOE application to the EPA to obtain a certificate of compliance with the 40 CFR 191 standard to open and operate WIPP. The EPA granted this certificate in 1998, and EPA oversight continues in periodic (i.e., every five years)

---

[1] For compliance with the standard of 40 CFR 191, the EPA issued rule 40 CFR 194 in 1996 to provide a regulatory interpretation of how these requirements would apply to WIPP.

recertifications. Changing some of the repository features (e.g., the design of the engineered seals to close underground rooms once they are filled with waste or the design of the seals to close the vertical shafts to the surface) would require regulatory approval because of their importance to the model of long-term performance.

DOE Management of TRU Waste

Transuranic (TRU) wastes are stored and managed at several DOE sites nationwide. To dispose of these wastes at WIPP, they must be retrieved from storage, characterized, repackaged (if necessary), and transported to WIPP, where they are unloaded from shipping containers and sent underground for emplacement in the disposal rooms.

These activities are conducted under the auspices of the National TRU Program administered by the DOE Carlsbad Area Office. DOE sites sending waste to WIPP must meet the waste characterization and transportation specifications that are contained in the WIPP waste acceptance criteria. The specifications on characterization and transportation operations are designed to meet all applicable regulations that have been promulgated by the EPA (chiefly through the Resource Conservation and Recovery Act, or RCRA), the U.S. Nuclear Regulatory Commission (USNRC), and the U.S. Department of Transportation (DOT). The waste characterization activities and the transportation system are described in more detail below.

Waste Characterization Activities

The characterization program described here has been developed for contact-handled[2] TRU waste and applied to date on non-mixed waste.[3] The methods, equipment, procedures, determination of uncertainty, and other protocols used at DOE sites to perform these characterizations are approved by both the DOE Carlsbad Area Office and the EPA. The major procedures are as described in the following sections:

Determination of the Origin and Composition of the Waste by Acceptable Knowledge. Acceptable knowledge of the origin and composition of the waste must be available in documentation to prove that the waste is of defense origin (by the terms of the Land Withdrawal Act, only defense-related TRU waste may legally be sent to WIPP) and to provide

---

[2]Contact-handled waste is that for which the maximum radiation dose rate at the surface of the waste container is less than 200 mrem per hour. Essentially no shielding other than the waste container is needed. Much of the DOE TRU waste has radioactivity due primarily to alpha-emitting actinides. Because alpha particles are relatively easy to shield, such waste would have a low surface dose rate and therefore would be classified as contact-handled waste.

[3]Mixed waste is waste with radioactive constituents regulated under the Atomic Energy Act mixed with hazardous chemical materials regulated under RCRA. Non-mixed radioactive waste is waste that can be shown not to contain RCRA-regulated materials.

characterization information on the waste constituents. The DOE Carlsbad Area Office and EPA use the acceptable knowledge documentation to certify each "waste stream" (i.e., waste-generating process), and TRU waste sent to WIPP must come from a certified waste stream.

Sampling and Analysis of Homogeneous Waste for RCRA Constituents. Most of the TRU waste is heterogeneous in nature and requires no further characterization beyond acceptable knowledge to satisfy the regulatory requirements of RCRA. For homogeneous waste, a fraction of the waste containers (e.g., 55-gallon drums or standard waste boxes) are cored to extract representative samples that are analyzed for constituents (e.g., volatile and semi-volatile organic compounds, toxic metals, and other hazardous chemicals) regulated by RCRA. Both the acceptable knowledge procedure (for heterogeneous waste) and the sampling and analysis procedure (for homogeneous waste) were proposed by DOE for the terms of operation that would be specified in its RCRA Part B permit. These terms have been accepted by New Mexico, which has authority delegated by the EPA to regulate RCRA materials and mixed waste and which issued the RCRA Part B permit in October 1999.

Real-Time Radiography. A real-time radiography procedure using x-rays is performed on all waste containers to look for items such as pressurized cans or free-standing liquids that are prohibited from being transported under DOT regulations. If any of these items are present in a waste container, its contents are repackaged, at which time the prohibited materials are removed. Another purpose of the radiography examination is to confirm the acceptable knowledge characterization information.

Visual Examination. A visual examination is performed on a fraction of the waste containers, by spilling the waste contents into a glovebox, to verify the acceptable knowledge and real-time radiography information. The value of this fraction was proposed by DOE to be two percent of the initial population of containers of each waste stream, and if these evaluations resulted in few miscertifications, then the percentage of subsequent waste containers to undergo visual examination would be reduced. In October 1999, New Mexico in its RCRA Part B permit stipulated the initial fraction of containers to undergo visual examination to be 11 percent.

Radioassay and Determination of Fissile Isotope Content. The number of curies of each transuranic isotope is determined by radioassay (e.g., gamma scans) to a specified precision and accuracy. The fissile isotope content is assessed using methods such as passive-active neutron systems. This information is used to ensure criticality safety, a USNRC requirement, which imposes a restriction on the amount (several hundred grams) of each fissile species per container. This restriction is less stringent than the amount derived from the gas generation model, discussed below.

Headspace Gas Sampling. Headspace gas sampling is carried out on all waste containers for flammable gases (specifically, volatile organic compounds, hydrogen, and methane). This procedure has been proposed as a means of checking on conformity with the DOT regulations (e.g., 40 CFR 173 and 40 CFR 177) and USNRC regulations (e.g., 10 CFR 71) that address the transport of flammable and/or gas-generating substances with radioactive materials (Mewhinney, 1998b). These regulations include the following statements:

- 49 CFR 173.21(g): "Packages which give off a flammable gas or vapor, released from a material not otherwise subject to this subchapter, likely to create a flammable mixture with air in a transport vehicle" are forbidden.
- 49 CFR 173.21(h): "Packages containing materials which will detonate in a fire" are forbidden.
- 49 CFR 173.24(b)(3): "There will be no mixture of gases or vapors in the package which could, through any credible spontaneous increase of heat or pressure, significantly reduce the effectiveness of the packaging."
- 49 CFR 177.848 specifies that flammable gases and radioactive materials "may not be loaded, transported, or stored together in the same transport vehicle or storage facility during the course of transportation unless separated in a manner that, in the event of leakage from packages under conditions normally incident to transportation, commingling of hazardous materials would not occur."
- 10 CFR 71.43(d): "A package must be made of materials and construction that assure that there will be no significant chemical, galvanic, or other reaction among the packaging components, among package contents, or between the packaging components and the package contents, including possible reaction resulting from in leakage of water, to the maximum credible extent. Account must be taken of the behavior of materials under irradiation."

DOE has proposed the headspace gas sampling procedure in its application to the USNRC for a licensing certificate on the transportation package (named the TRansUranic PACkage Transporter, or TRUPACT-II) that is loaded with waste containers for transport by truck to WIPP.

Repackaging of Waste to Meet Wattage Limits Imposed by a Radiolytic Gas Generation Model. Gas generation can occur during the transport of a waste container to WIPP. The radiolytic generation of hydrogen gas in TRU waste comes from the co-disposal of organic materials (containing hydrogen) with alpha-emitting radionuclides, which irradiate the organic matter to produce H^+ ions that combine to form H_2 molecules. The current gas generation model is based on assumptions about the configuration of organic materials and radionuclides. It relates the concentration of hydrogen gas in any headspace to the alpha activity (i.e., activity from alpha-emitting radionuclides) within each waste container. More than one gaseous headspace can exist in a waste container, primarily because TRU waste, when generated and disposed in DOE facilities, was contained within layers of confinement provided by plastic bags that may still be intact and thereby inhibit the flow of hydrogen.

By placing a 5 percent (mole fraction) limit on the maximum H_2 concentration within any headspace, this gas generation model calculates an upper limit, commonly expressed as a maximum thermal wattage, on the alpha activity allowed for the entire waste container. These wattage limits are a function of the waste materials and the number of layers of confinement provided by plastic bags. Because of its conservatism, the value of 5 percent H_2 (as a mole fraction) in air as a "flammability limit" can be used in any USNRC license application for a transportation package without the need for further safety analysis.

For example, for a 55-gallon drum containing a plastic liner and heterogeneous debris with plutonium inside three layers of sealed plastic bags, the wattage limit is approximately 0.028 W (DOE, 1996b, p. 5-6e), which corresponds to a limit of 14 g (0.89 Ci) of plutonium-239 or 0.049 g (0.84 Ci) of plutonium-238. Waste containers containing more wattage than the maximum value allowed by the model have their waste contents repackaged to distribute the TRU waste into configurations that will meet these wattage limits. This is accomplished by spilling these contents into shielded gloveboxes and dividing the waste into several new containers, each filled with a fraction of the contents of the original waste container. At Los Alamos National Laboratory in 1998-1999, gas generation restrictions resulted in the repackaging of 36 drums of plutonium-238 waste from the waste stream "TA-55-43" into approximately 120 drums that were placed inside standard waste boxes.[4]

The output of the characterization program is a set of characterization data for each waste container. If the characterization information is within acceptable limits as determined by the waste acceptance criteria and quality assurance program plan (or waste analysis plan) specifications, the waste container is certified and approved for shipment to WIPP.

Truck Transportation to WIPP

At the DOE sites containing TRU waste, the certified TRU waste containers are loaded inside TRUPACT-II shipping containers that are then sealed with a vacuum-tight seal. The TRUPACT-II is classified and regulated as a "Type B" package for fissile materials.[5] To ensure that the waste contents are safely contained during normal shipment conditions and accident scenarios, this transportation package must meet design features such as double containment (i.e., it must have an inner and outer container) and a vacuum seal. Within the inner container, two standard waste boxes, fourteen 55-gallon drums, or one standard waste box and seven 55-gallon drums can be placed. These waste containers are loaded into the TRUPACT-II using an overhead crane in a bay of a building that a truck can drive into to avoid the need to unfasten the TRUPACT-II from the trailer.

---

[4] A 55-gallon drum has a volume of approximately 0.2 m^3, whereas a standard waste box is a 1.9 m^3 container that can hold three 55-gallon drums.

[5] This designation is a regulatory term to designate packages used to transport plutonium isotopes, which are contained in TRU waste.

The trucks travel to WIPP on approved highway routes during approved times and maintain communication with a DOE control center. In addition to a cellular telephone and a citizens band radio, each truck contains a satellite transponder that enables it to be tracked en route using DOE's satellite-based telecommunications system, the TRANSportation Tracking and COMmunication (TRANSCOM) System. The TRUPACT-IIs are inspected at the WIPP site and their contents (waste-filled drums or boxes) are unloaded and delivered to an underground elevator for emplacement into rooms excavated in the subsurface salt bed.

Appendix B

Joint USNRC and EPA Guidance on Mixed Waste

A joint U.S. Nuclear Regulatory Commission (USNRC) and U.S. Environmental Protection Agency (EPA) document (62 FR 62079, 1997) provides regulatory guidance outlining the testing requirements for mixed radioactive and hazardous waste. In this dual agency guidance document, the EPA and USNRC position is that a combination of common sense, modified sampling procedures, and cooperation between state and federal regulatory agencies will minimize any hazards associated with sampling and testing mixed waste.

Waste generators may determine whether their waste is a Resource Conservation and Recovery Act (RCRA) hazardous waste based on knowledge of the materials or chemical processes that were used. That is, RCRA regulations do not require testing of the waste.

Therefore, where sufficient knowledge of materials or of the process exists, the generator need not test the waste to determine that it possesses a hazardous characteristic, which would necessitate that RCRA be applied (although generators and subsequent handlers would be in violation of RCRA if they managed hazardous waste erroneously classified as nonhazardous outside the RCRA hazardous waste system). For this reason, facilities wishing to minimize testing often assume that a questionable waste is hazardous and handle it accordingly.

Flexibility exists in the hazardous waste regulations for generators; operators of treatment, storage, and disposal facilities; and mixed waste permit writers to tailor mixed waste sampling and analysis programs to address radiation hazards. For example, upon the request of a generator, a person preparing a RCRA permit for such a facility has the flexibility to minimize the frequency of mixed waste testing by specifying a low testing frequency in a facility's waste analysis plan. The EPA position, as stated in 55 FR 22669 (1990), is that the frequency of testing is best determined on a case-by-case basis by the permit writer.

The joint USNRC-EPA agency guidance document (62 FR 62079, 1997) appears to the committee to provide appropriate guidelines for implementation and integration of RCRA requirements for mixed TRU waste. Implementation of this regulatory guidance could significantly reduce the testing protocols and associated radiation exposure of personnel. At present, the procedures specified in the waste acceptance criteria and quality assurance program plan documents and in the RCRA Part B permit for the testing of mixed waste seem at odds with the ALARA (as low as reasonably achievable) principle.

Appendix C

Biographical Sketches of Committee Members

B. John Garrick, *Chair,* independent consultant, is a co-founder of PLG, Inc., an international engineering, applied science, and management consulting firm in Newport Beach, California. He received his B.S. degree from Brigham Young University and his M.S. and Ph.D. degrees in engineering and applied science from the University of California, Los Angeles. His professional interests involve risk assessment in applications in fields such as nuclear energy, space, and defense, and in the chemical, petroleum, and transportation industries. He has received numerous awards, including the Society for Risk Analysis Distinguished Achievement Award. He was appointed to the U. S. Nuclear Regulatory Commission's Advisory Committee on Nuclear Waste in 1994, for which he is now Chairman. Dr. Garrick was elected to the National Academy of Engineering in 1993. He has been a member of the Committee on the Waste Isolation Pilot Plant since 1989.

Mark Abkowitz, professor of civil engineering at Vanderbilt University and director of the Center for Environmental Management Studies, has many years of experience in hazardous materials transport. He has published widely on transportation issues such as the risks of transporting high-level radioactive waste. He is a member and former chairman of the NRC Transportation Research Board standing committee on hazardous materials transport.

Alfred W. Grella, independent nuclear and hazardous materials transportation consultant, retired in 1990 from a career in U.S. government service, first at the Department of Transportation and later at the U.S. Nuclear Regulatory Commission. His distinguished career spans 40 years as a professional in health physics, health protection, transportation, inspection and enforcement, training, and related regulatory activities. Mr. Grella received a Bachelor's degree in chemistry from the University of Connecticut and completed the one-year management program at the National Defense University Industrial College of the Armed Forces. He has authored over 30 published papers. He is a member of the American Nuclear Society and a Fellow of the Health Physics Society. Mr. Grella received the M. Sacid (Sarge) Ozker Award in 1996 for distinguished service and eminent achievement in the field of radioactive waste management.

Michael Hardy, president of Agapito Associates, Inc., has experience in numerical modeling and field experimentation in practical, engineering-oriented studies to gather characterization data and to evaluate the merits of design features of proposed high-level waste repositories. Dr. Hardy is a member of the Society of Mining, Metallurgical and Exploration Engineers, Inc., and the American Society of Civil Engineers (ASCE). He is Chairman of the Underground Technical Research Council, a joint ASCE/American Institute of Mining, Metallurgical, and Petroleum Engineers Committee.

Stanley Kaplan, principal of Kaplan & Associates, Inc., is one of the early practitioners of the discipline now known as Quantitative Risk Assessment and a major contributor to its theory, language, philosophy and methodology. Dr. Kaplan is a Fellow of the Society for Risk Analysis and the author of a number of the seminal papers in this field. He is one of the first contributors to the Russian science TRIZ, the Theory of the Solution of Inventive Problems, and currently consults and teaches in this area. He is a founder and board chairman of Bayesian Systems, Inc., a Washington-based company developing diagnostic, decision, simulation, and business management software. Dr. Kaplan is the recipient of several awards and honors, including the Society for Risk Analysis Distinguished Achievement Award in 1996. Dr. Kaplan was elected to the National Academy of Engineering in 1999.

Howard M. 'Skip' Kingston is professor of chemistry in the Department of Chemistry and Biochemistry and in the Center for Environmental Research and Education. Also at Duquesne University, he is director of the Center for Microwave and Analytical Chemistry. His research interests include the development, automation, and standard encapsulation and transfer of analytical analysis methods. For the past several years, he has been actively involved in advancing the area of microwave sample preparation through basic research and the development of procedures that have been adopted by the EPA as standard methods. From 1976 to 1991 he was a supervisory research chemist in the Inorganic Analytical Research Division of the National Institute of Standards and Technology (NIST), where he conceived and managed the Consortium on Automated Analytical Laboratory Systems dedicated to developing automated analytical capability for industry. He has received numerous awards for his pioneering work in several areas, including R&D 100 Awards in 1996 and 1998, the IR 100 Award in 1987, the 1988 "Pioneer in Laboratory Robotics" award, the 1990 NIST Applied Research Award, the Department of Commerce Bronze Medal in 1990, the Award of Merit from the Federal Laboratory Consortium in 1991, and the EPA RCRA Service to Others Award in 1998. He has co-edited and co-authored the American Chemical Society professional reference texts *Introduction to Microwave Sample Preparation: Theory and Practice* (1988) and *Microwave Enhanced Chemistry: Fundamentals, Sample Preparation, and Applications* (1997).

He holds multiple patents in the field of speciation, microwave chemistry, and chelation chromatography.

W. John Lee, Peterson Chair and professor of petroleum engineering at Texas A&M University and formerly executive vice-president of technology at S. A. Holditch & Associates, Inc., has expertise in petroleum reservoir imaging, flow tests in low-permeability formations, and enhanced recovery practices. Professor Lee was elected to the National Academy of Engineering in 1993.

Milton Levenson, independent consultant, is a chemical engineer with over 50 years of experience in nuclear energy and related fields. His technical experience includes work in nuclear safety, fuel cycle, water reactor technology, advanced reactor technology, remote control technology, and sodium reactor technology. His professional experience includes research and operations positions at Oak Ridge National Laboratory, Argonne National Laboratory, the Electric Power Research Institute, and Bechtel. Mr. Levenson is the past president of the American Nuclear Society; a fellow of the American Nuclear Society and the American Institute of Chemical Engineers; and the recipient of the American Institute of Chemical Engineers' Robert E. Wilson Award. He is the author of over 150 publications and presentations and holds three U.S. patents. He received his B.Ch.E. from the University of Minnesota. He was elected to the National Academy of Engineering in 1976.

Werner F. Lutze, professor of chemical and nuclear engineering at the University of New Mexico and director of the UNM Center for Radioactive Waste Management (CeRaM), has over 25 years of research experience in materials science and geochemical issues relevant to the management of radioactive wastes, including selective mineral ion-exchange processes, repository near-field chemistry, waste form development, and trace analyses. He has published widely on weapons plutonium immobilization, waste disposal, and the chemistry of nuclear materials. Professor Lutze is a member of several professional organizations, including the Materials Research Society, the German Nuclear Society, and Sigma Xi.

Kimberly Ogden, associate professor of chemical and environmental engineering at the University of Arizona, has conducted research with Los Alamos National Laboratory collaborators to design treatment methods for remediating hazardous waste sites containing both toxic metals and organics, including plutonium-cellulose mixtures. She is also engaged in collaborations with ECO Compliance Inc. in preparing proposals and reports for the remediation of hazardous waste sites. Professor Ogden has authored or co-authored several book chapters, papers, and presentations in environmental science and technology. She is a member of the American Institute of Chemical Engineers, the American Association for the Advancement of Science, and the American Chemical Society.

Martha Scott, associate professor of oceanography at Texas A&M University, is a researcher in marine radiochemistry and geochemistry. Her present research involves radionuclide distribution in the Russian Arctic. Her work has dealt with the interaction between oceans and rivers, transport of materials in the marine environment, and chemistry of manganese nodules. The behavior of plutonium isotopes in rivers, estuaries, and marine sediments has been one of her longstanding research interests. She served for two years as an associate program director for chemical oceanography at the National Science Foundation (1992-1993). She received the Ph.D. degree from Rice University and was a National Science Foundation post doctoral fellow at Yale University.

John M. Sharp, Chevron Centennial Professor of Geology at The University of Texas at Austin, leads an active research program in hydrology. Professor Sharp has authored and co-authored over 200 journal articles, books, reports, and presentations. He is a fellow of the Geological Society of America and recipient of its O.E. Meinzer award (1979) and the American Institute of Hydrology's C.V. Theis Award (1996). Dr. Sharp is the current editor of *Environmental and Engineering Geoscience*. He received his B. Geological E. with Distinction from the University of Minnesota and his M.S. and Ph.D. degrees in Geology from the University of Illinois.

Paul G. Shewmon, emeritus professor of materials science and engineering at the Ohio State University, received a B.S. degree in metallurgical engineering from the University of Illinois and M.S. and Ph.D. degrees, also in metallurgical engineering, from the Carnegie Institute of Technology. He recently retired as Humbolt Senior Scientist at the Max Planck Institute Metallforschung in Stuttgart. He has received the ASM deMille Campbell Lecture and Award and the TMS Institute of Metals Lecture & Mehl Medal. He was elected to the National Academy of Engineering in 1979.

James Watson, Jr., professor of environmental sciences and engineering and the Director of the Air, Radiation, and Industrial Hygiene Program at the University of North Carolina at Chapel Hill, holds an M.S. degree in physics from North Carolina State University and a Ph.D in environmental sciences and engineering from the University of North Carolina at Chapel Hill. Professor Watson is accomplished in the fields of environmental radioactivity and radioactive waste management. He has received the Underwood and McGavran Awards for excellence in teaching and the Greenberg Alumni Endowment Award for excellence in teaching, research, and service. He is a past president of the Health Physics Society and a past chairman of the Radiological Health Section of the American Public Health Association. He has served on the Environmental Protection Agency's Radiation Advisory Committee and the executive committee of the agency's Science Advisory Board. He is a past chairman of the North Carolina Radiation Protection Commission and currently chairs the commission's Committee on Low-Level Radioactive Waste Management.

Ching H. Yew, an independent consultant and emeritus professor from The University of Texas at Austin, has specialized in the study of hydraulic fracturing and borehole stability. Dr. Yew is a fellow of the America Society of Mechanical Engineers and a member of the Society of Petroleum

Engineers. Dr. Yew has authored a text and published several articles concerning hydraulic fracturing and borehole stability. The computer code developed by him has been adopted for field use by many oil and gas industries.

Appendix D

Acronyms

ALARA	as low as reasonably achievable
ATMX	Atchison Topeka Munitions private railcar
CFR	Code of the Federal Regulations
CEMRC	Carlsbad Environmental Monitoring Research Center
DOE	U.S. Department of Energy
DOT	U.S. Department of Transportation
EPA	U.S. Environmental Protection Agency
NORM	naturally occurring radioactive material
NRC	National Research Council
RCRA	Resource Conservation and Recovery Act
TRANSCOM	TRANSportation Tracking and COMmunication system
TRU	transuranic
TRUPACT	TRansUranic PACkage Transporter
USNRC	U.S. Nuclear Regulatory Commission
WIPP	Waste Isolation Pilot Plant

Appendix A2

DOE's Response to the Interim Report

Department of Energy
Carlsbad Field Office
P. O. Box 3090
Carlsbad, New Mexico 88221
October 30, 2000

National Research Council
Committee on the Waste Isolation Pilot Plant
Board on Radioactive Waste Management
C/O Kevin D. Crowley, Director
2101 Constitution Avenue, NW
Washington, DC 20418

Dear Committee members:

Thank you for your excellent work on the Waste Isolation Pilot Plant Interim Report. We have read your report with great interest and agree with its principles. I am attaching a response to each of the recommendations which you may wish to consider in developing your Final Report.

To briefly summarize, our responses are as follows:

- Data on NORM is being collected and a database developed.
- We are actively pursuing reduction and elimination of self-imposed waste characterization requirements that lack a technical or safety basis.
- We are working closely with the Nuclear Regulatory Commission to minimize the impact of the 5% hydrogen limit, and to reduce or eliminate unnecessary repackaging of waste.
- TRANSCOM has been completely revised to include requested updates and to incorporate specific WIPP requirements.
- We are working with the states to identify and remedy gaps in emergency response coverage.

In conclusion, your recommendations have been adopted as the cornerstone of our planning to bring WIPP to its full potential as the solution to managing our nation's TRU waste.

Sincerely,

Dr. Inés Triay, Manager

Attachments
Additional copies of Rail Study
    sent under separate cover

cc: Lynne Wade, EM 23
    Matthew Silva, EEG

CBFO:OOM:IRT:KJB 00-0467 UFC #5480

Response to National Research Council Recommendations
Committee on the Waste Isolation Pilot Plant – Interim Report

I. NORM in the WIPP Vicinity

Recommendation: "DOE should develop and implement a plan to sample oil-field brines, petroleum, and solids associated with current hydrocarbon production to assess the magnitude and variability of naturally occurring radioactive material (NORM) in the vicinity of the WIPP site."

Response: DOE agrees with this recommendation. The New Mexico State University Carlsbad Environmental Monitoring & Research Center (CEMRC) has undertaken a project to carry out the recommended assessment, as part of CEMRC's ongoing WIPP Environment Monitoring project. Although analyses of certain naturally occurring radioactive materials in hydrocarbon and scale matrices are somewhat standardized, the sensitivity of the standard methods will likely not be acceptable for at least a portion of the matrices targeted, resulting in the need for method enhancement research. In addition, no published methods are available for analyses of plutonium, americium and other TRU components of concern in hydrocarbons, so this analytical task will require extensive method development and validation prior to initiation of analyses of actual samples.

NORM is an extremely sensitive topic in the oil and gas production industry in the region of the WIPP. As an example, one major exploration and production company that operates in Lea and Eddy counties recently submitted 48 pages of critique on draft Regulation Guidelines for the Management of NORM in the Oil and Gas Industry in New Mexico, which was issued by the New Mexico Environment Department in 1999. Initial contacts with those familiar with local and regional companies indicate that it is likely that many if not most operators will decline to cooperate. To create the maximum likelihood of obtaining cooperation, an option for anonymity will be offered to the operators, using a form of "double-blind" identification. This system would involve collection of samples by a commercial third-party service company that is acceptable to the operator, submittal of the samples to CEMRC without identification of the operator or well location (formation and production pool only), resulting in CEMRC reporting of results without specific operator or well identification.

Path Forward: CEMRC received approval in August 2000, from DOE to proceed with the project without direct involvement of DOE in contacting affected production operators. A plan for a study entitled "Characterization of radioactive elements in oil and gas production in the vicinity of the WIPP" was developed by CEMRC. The initial phases of the study are in progress, including completion of a database of active wells and operators, development of sample collection and handling plans, and identification of commercial sample collection services currently operating in the area. Initial contacts with operators to solicit participation in the study will occur during November through February 2000. Contingent on cooperation of enough operators to create a representative sampling design, sample collection would be conducted during March through August 2001.

-2-
Response to National Research Council Recommendations
Committee on the Waste Isolation Pilot Plant – Interim Report

II. Waste Characterization

Recommendation: "DOE should eliminate self-imposed waste characterization requirements that lack a legal or safety basis."

Response: DOE agrees with this recommendation. DOE has developed and begun the implementation of a strategy to systematically improve the Waste Analysis Plan by reducing the frequency of waste characterization and implementing methods that make characterization simpler, less expensive and, above all, safer.

On August 8, 2000, the New Mexico Environment Department approved two packages of Class 2 modifications to the WIPP's Hazardous Waste Facility Permit. These two packages include three requests submitted on April 5, 2000 and three submitted April 20, 2000.

Approval of these modifications results in significant cost savings associated with waste characterization and will reduce radiation exposures to workers. A summary of the approved modifications follows:

- The "miscertification rate" of TRU waste was revised to apply to the waste summary category group instead of each waste stream. This results in a ten-fold reduction in number of drums that must be opened for visual examination (VE).

- The solids sampling requirements for analysis of VOCs have been revised to allow use of one subsample instead of three subsamples. This will avoid a cost of approximately ten million dollars that INEEL would have had to spend in re-analyzing the samples.

- The number of headspace gas samples required has been reduced for two types of waste streams to a statistically selected number of drums, instead of 100% sampling. The two types of waste streams now eligible for statistical headspace gas sampling are wastes that have been thermally processed and homogeneous wastes with "acceptable knowledge" that demonstrates no volatile organic compounds have been present in the waste.

Several modifications have been prepared and submitted that specifically address safety issues associated with TRU waste handling and disposal. One such modification, submitted in October 2000, will allow generators to remove from consideration for VE any containers that pose a safety concern. For example, if a generator determines that opening a container with a high fissile gram content is a safety hazard, that container can be ruled ineligible for VE and another container selected.

-3-

Path Forward: The next modification request, which will be submitted in November 2000, will provide alternatives to VE as a quality control check on radiography. Should the modification be granted, DOE intends to implement this change across the complex.

The Permit modification requesting authorization for remote handled waste disposal at WIPP (to be submitted in December 2000) presents a performance based waste analysis plan that emphasizes the use of nonintrusive characterization techniques and eliminates the need for headspace gas sampling and analysis, solids sampling and analysis, VE, and other confirmatory measurements.

The DOE also plans to collect data from waste characterization activities that will allow the systematic reduction or elimination of headspace gas sampling, solids sampling, VE, and radiography. These changes will be promptly implemented as suitable supporting data are identified.

-4-
Response to National Research Council Recommendations
Committee on the Waste Isolation Pilot Plant – Interim Report

III.A Derive A More Realistic Gas Generation Model

Recommendation: "DOE should derive a more realistic radiolytic gas generation model, validate it through confirmatory testing, use the results to recalculate gas generation limits, and seek regulatory approval to implement these limits."

Response: DOE agrees with this recommendation. An application for Revision 19 of the TRUPACT-II Safety Analysis Report was submitted to the Nuclear Regulatory Commission in April 2000. Among other things, the application includes the following:

- Matrix Depletion – The g-values of organic materials decline as a function of absorbed radiation dose. Testing performed at Los Alamos National Laboratory demonstrated that the g-value of polyethylene declines from an initial g-value of 3.4 to 1.1. The application requests the use of the lower g-value. When approved this new g-value will increase the allowable wattage up to a factor of 3 (depending on the packaging configuration).

- Option for the mixing of shipping categories that will allow the sites to ship payloads with different waste forms and to take credit for the use of dunnage containers (additional void volume and reduced gas concentrations).

- Use of more realistic g-values to take credit for non-gas generating materials present in the waste, based on percentages of moisture or organic material present. The previous model assumed a worst-case, 100% moisture/organic material scenario. (This change has been approved by the Nuclear Regulatory Commission.)

- Use of a new shipping category notation that accurately reflects the packaging configuration of the waste. The previous notation grouped all sites under selected worst-case packaging configurations. (This change has been approved by the Nuclear Regulatory Commission.)

Path Forward: The application for Revision 19 was submitted to the Nuclear Regulatory Commission in April 2000 and is scheduled for approval in January 2001. We believe this to be a very responsive review cycle. Taken as a whole, Revisions 17, 18, and 19 provide an increase of up to 100 times the wattage that was allowed under Revision 16.

-5-
Response to National Research Council Recommendations
Committee on the Waste Isolation Pilot Plant – Interim Report

III.B Safety Analysis to Determine the Damaging Concentration of Hydrogen

Recommendation: "DOE should perform a safety analysis to determine the concentration and quantity of hydrogen that, upon ignition, could damage the seals of the TRUPACT-II shipping container. The goal of the safety analysis would be to demonstrate whether such an event could occur inside a waste package, and whether the energy associated with such an event could result in the rupture of containment provided by the TRUPACT-II. This analysis could provide the rationale to obtain relief from the 5 percent hydrogen flammability limit and should form the basis for a future modification to the present TRUPACT-II license."

Response: DOE agrees with the recommendation. Performing the safety analysis, which may include testing, would be the first step toward an application. Preliminary review of the recommendation has raised an issue of handling drums at the WIPP that could have a potentially flammable gas mixture. The safety analysis should be extended to waste handling operations at WIPP. If the safety analysis indicates that there is not a safety concern, then an application would be submitted to the Nuclear Regulatory Commission for their review and approval. One proposed solution is to encapsulate the waste in a manner that would contain any detonation that might occur. It is noted that there is no precedence for Nuclear Regulatory Commission approval of shipment of a flammable gas and radioactive material in the same package. Also, the U.S. Department of Transportation (DOT) prohibition against shipping containers of flammable gas and radioactive material on the same vehicle if they could co-mingle would have to be addressed.

Path Forward: The following steps will be pursued to respond to this recommendation:

- Perform a safety analysis to determine whether WIPP could unload drums of waste that contained flammable gas.

- Assess ArrowPAK^{tm} suitability for macro-encapsulation to contain potential deflagration events.

- Determine the incremental quantity of waste that could benefit from implementation of this recommendation (assuming the current application for Revision 19 to the TRUPACT-II Safety Analysis Report is approved).

- Perform the recommended analysis and/or testing.

- Prepare an application and submit to the Nuclear Regulatory Commission for review.

- Seek DOT concurrence.

-6-
Response to National Research Council Recommendations
Committee on the Waste Isolation Pilot Plant – Interim Report

III.C Technical Approaches to Reduce Hazards Such As Inert Gas

Recommendation: "DOE should consider technical approaches for reducing hazards from hydrogen generation, such as filling the headspace of the waste containers or the shipping containers with an inert gas to displace air and thereby reduce the flammability hazard."

Response: DOE agrees with this recommendation and is actively pursuing several alternative technologies such as hydrogen "getters." There are several technical issues associated with this recommendation that would have to be investigated:

- Would a drum containing multiple layers of confinement around the TRU waste benefit from the proposed technology; e.g., inert gas in the drum headspace?

- Does the proposed technology require opening individual payload containers or does it apply to the TRUPACT-II; i.e. inert the TRUPACT-II ICV headspace?

- Does the proposed technology prevent or mitigate detonation/deflagration inside multiple layers of confinement or inside the TRUPACT-II? (See recommendation III.B above.)

A method of measuring the flammable gas concentration in the headspace of a 55-gallon drum has been included in the application for Revision 19 of the TRUPACT-II Safety Analysis Report. This only requires a single measurement that can be made in real-time instead of a lengthy gas generation rate test that requires measuring the rate of change of hydrogen gas over several hours or days.

Path Forward: The following steps will be pursued to respond to this recommendation:

- Alternative technical recommendations such as inert gas or hydrogen getters will be considered and a report prepared.

- Perform analysis and/or testing of those alternative technologies that look promising.

- Submit an application to the NRC for any technologies that can be supported by analysis or test results.

-7-
Response to National Research Council Recommendations
Committee on the Waste Isolation Pilot Plant – Interim Report

III.D Reevaluate the Feasibility of Rail

Recommendation: "DOE should reevaluate the technical and regulatory feasibility of shipping high-wattage TRU waste using ATMX railcar shipping system."

Response: DOE agrees with the recommendation to reconsider the use of rail. DOE is currently reviewing a recently commissioned rail study report (included as an attachment). The report concluded that shipment of TRUPACT-II by rail is not cost effective unless significantly reduced rail rates are available. The report also recommended investigation of a new shipping package (TRUPACT-III) for shipping high wattage waste and oversize boxes by rail or truck.

DOE has not made a decision regarding the use of ATMX railcar for shipments to WIPP. Using ATMX would require one of the following to occur:

1. Approval by the NRC – this would require exemption(s) from several of the requirements in 10 CFR 71.

    Or,

2. Revision of the Consultation and Cooperation Agreement between the State of New Mexico and DOE, plus revision to the WIPP Land Withdrawal Act.

Path Forward: The following steps will be pursued to respond to this recommendation:

- Determine the incremental inventory of TRU waste that could benefit from rail shipment (assuming the current application for Revision 19 to the TRUPACT-II Safety Analysis Report is approved).

- Evaluate and compare the benefits and regulatory difficulty of two options – TRUPACT-III vs. ATMX.

- Make a decision based on information obtained.

- Proceed with the chosen option.

-8-
Response to National Research Council Recommendations
Committee on the Waste Isolation Pilot Plant – Interim Report

IV. DOE's Communication and Notification Program

Recommendation: DOE should consider cost-effective ways to improve the reliability and ease of use of the TRANSCOM system, either by improving or replacing it utilizing current technologies, and ensure the future system meets the WIPP and other stakeholders needs.

Response: DOE agrees with the recommendations and, in fact, has been working toward this effort since mid-1998. On September 12, 2000 in Albuquerque NM, the DOE National Transportation Program-Albuquerque (NTPA) presented a response to the Interim Report recommendations. Committee members represented were Dr. Mark Abkowitz and Mr. Al Gruella. The presentation included improvements to the present TRANSCOM system and a demonstration of the new web-based TRANSCOM2000 system.

- NTPA has identified the problems and has provided resolution that has increased reliability and stability of the present client server Windows-based system for over 7 months. These problems included difficulty in logging in, extended download times, loss of positional data, and date and time anomalies. Feedback from system users indicates that the system provides consistently reliable and accurate information, is more user friendly, and is meeting the needs of our customers. Customer complaints related to using the client-server version have been reduced to a very rare occurrence.

- In May 1999, NTPA hosted a TRANSCOM user application design session to team with DOE, State and Tribal TRANSCOM users, to develop system requirements for a new Internet-based communications and tracking system. In August 2000, the new application was beta-tested and discrepancies were identified and addressed.

- TRANSCOM2000 uses various commercial state-of-the-art Internet compatible software elements. These include: Object F/X GIS mapping software and engine, QTRACS satellite positional and two-way communications software, Oracle Relational DBMS 8I, Oracle Report and Oracle Forms 6I. The TRANSCOM Communications Center will have up-to-the-minute satellite weather service available.

- Security elements on the new application include native Oracle encryption, operation on the Secure Socket Layer (SSL), and multiple layers of application access security down to the database level. A firewall will also reside between the public and the application server.

-9-
- Positional update information will be received via frame-relay between the satellite service provider and the TRANSCOM2000 database. These positional updates will be requested every 2 minutes and should be available to the users on average of 2-5 minute intervals.

Path Forward: The estimated schedule for implementation of the TRANSCOM2000 is mentioned below. Firewall configuration and connectivity to the WIPP has been established. The WIPP Central Monitoring Room operators will be trained during the initial implementation process.

- The Major Application Security and Test Plans are under development. These plans must be approved prior to implementation per DOE Order. Completion date: November 2000.

- DOE/AL Operations configuration: in process. Weather and Qtracs servers being installed, firewall server built and awaiting software installation, Operations Center fully staffed. Completion Date: November/December 2000.

- Demonstration of operational readiness: December 2000.

-10-
Response to National Research Council Recommendations
Committee on the Waste Isolation Pilot Plant – Interim Report

V. DOE's Emergency Response Program

Recommendation: The committee recommends that DOE explore with states and other interested parties how to develop processes and tools for maintaining up-to-date spatial information on the location, capabilities, and contact information of responders, medical facilities, recovery equipment, regional response teams, and other resources that might be needed to respond to a WIPP transportation incident. This assessment should explore which organization(s) should develop and maintain the capability to generate and maintain such information. DOE should also determine where emergency response capability is currently lacking, identify organization(s) responsible for addressing these deficiencies, and take action to address them.

Response: The information that needs to be gathered and analyzed must come from the state, tribal, and local governments. According to preliminary telephone calls to the Western Governors' Association and the Federal Emergency Management Agency, there is not a national or state system that currently tracks the information that the Committee recommends the DOE collect and analyze. DOE must determine what level of participation the state, tribal and local governments are willing to have in the collection of this information and the maintaining of a database or reports. Logistically, the DOE will be trying to collect information from 30 states and 12 tribal governments which translates to data on over 100,000 emergency responders, and thousands of fire and police departments, ambulance services, and hospitals.

Path Forward: DOE will send letters to the regional, state, and Indian Tribal governments with whom it has cooperative agreements, asking them to communicate the recommendation of the Committee to all of its affected members. They will be asked to analyze their current data collections systems, and to define their willingness to participate in a regular assessment as recommended by the Committee.

They will further be asked to define their funding and manpower requirements, to submit the required data, and make counter recommendations that may fulfill the intent of the recommendation. This would include their recommendations on where the information is to be maintained and who will have the responsibility to analyze and make recommendations for improvement based on that data.

Appendix B

Human Intrusion Scenarios

Oil, gas and other mineral resources are frequently found in association with salt beds, such as the Salado, where the WIPP is situated. The region around the WIPP has known a high rate of drilling activities in the past and future energy trends indicate that there will be incentives to explore the region again, once institutional controls are removed (starting 100 years after the closure of the repository). The risk of drilling directly into the repository and thus creating pathways for the release of radionuclides into the environment will then increase. Drilling through the repository could transport radioactive materials from the repository to the surface or bring water in contact with substances stored in the repository. The following two scenarios are possible sources of concern about the performance of the repository and have been taken into account in the performance assessment of the WIPP.

1. If there were an oilfield water-flooding operation in the vicinity of WIPP, a large amount of brine could flow from a leaky injection well and induce a hydraulic fracture in the anhydrite (or marker bed) directly above or below the WIPP repository (Box B.1). If, at some later time, another well were drilled through the repository and into this brine-filled fracture, the high-pressure brine in the fracture could flow through the borehole and flood the repository causing a release of radioactive materials. The scenario is known as the Hartman scenario.

2. Direct drilling into the WIPP repository could allow circulating drilling fluid to bring radioactive materials to the surface through a borehole as cuttings or spallings. The situation could be serious if the repository were flooded with high-pressure brines. The Sandia National Laboratories examined three possible flooding scenarios, designated as E1, E2, and E1E2, in their performance assessment. These scenarios are briefly explained in Boxes B.2, B.3, and B.4 and they are described in detail in the Compliance Certification Application (DOE, 1996).

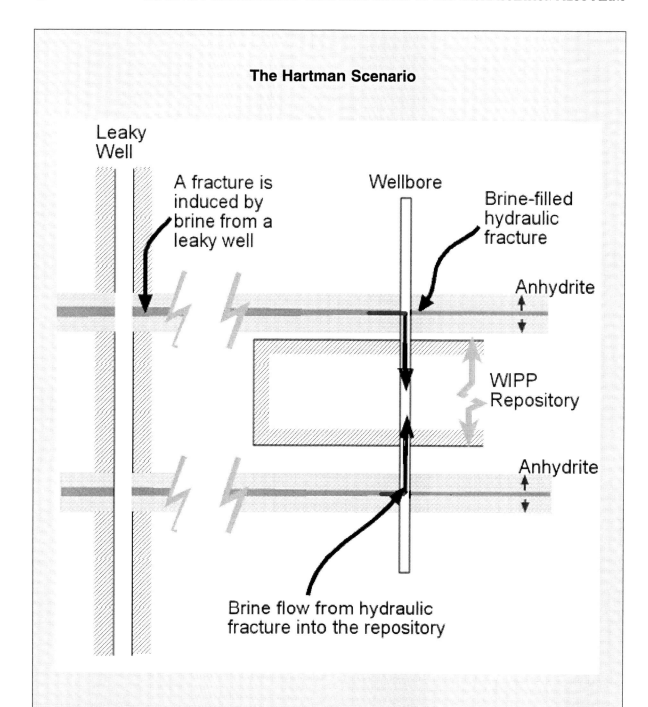

Box B.1: The Hartman Scenario is a scenario in which water from a leaky injection well induces a hydraulic fracture in the anhydrite below or above the repository. If, at some later time, another well is drilled through the repository, the water in the fracture could flow through borehole into the repository.

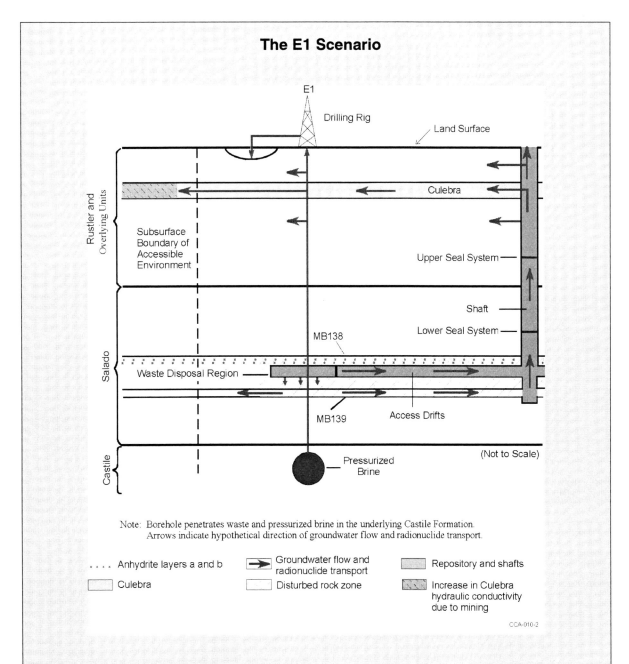

Box B.2: The E1 Scenario is any inadvertent penetration of a waste panel by a borehole that also penetrates a Castile brine reservoir. Sources of brine in the E1 scenario are the brine reservoir, the Salado, and under certain conditions, the units above the Salado.

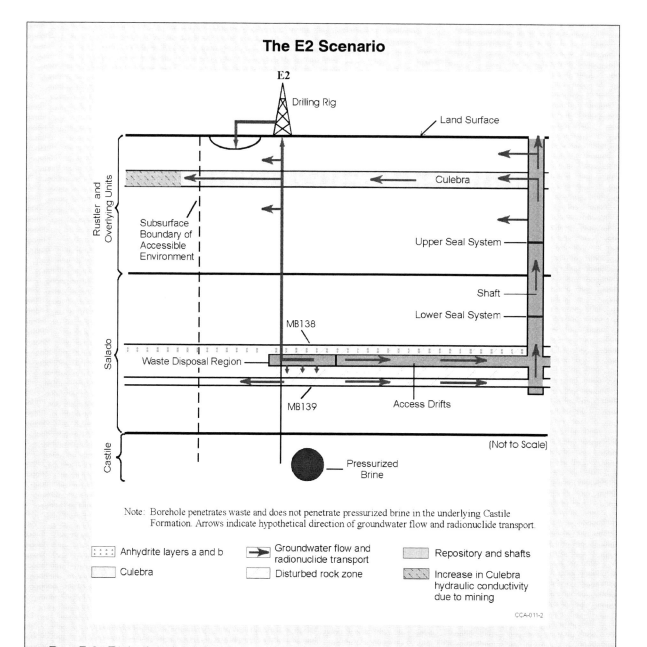

Box B.3: E2 is the simplest scenario for inadvertent human intrusion into a waste disposal panel. In this scenario, a panel is penetrated by a drill bit; cuttings, caving, spallings, and brine flow releases may occur in the borehole after it is plugged and abandoned. Cuttings will be discharged at the surface and may contain waste material if the borehole penetrates waste drums. *Cavings*, which include material eroded from the borehole wall during drilling, may also contain radionuclide waste from the repository horizon. Spallings include solid material carried into the borehole during rapid depressurization of the waste disposal region. The repository horizon could be pressurized by gas generation from degradation of the waste, organic materials and metal corrosion. Brine can be present in the Salado from natural sources or human activities associated with other drilling or production activities. Release to the biosphere is either to the surface or through the Culebra via a leaking *casing*.

APPENDIX B: HUMAN INTRUSION SCENARIOS

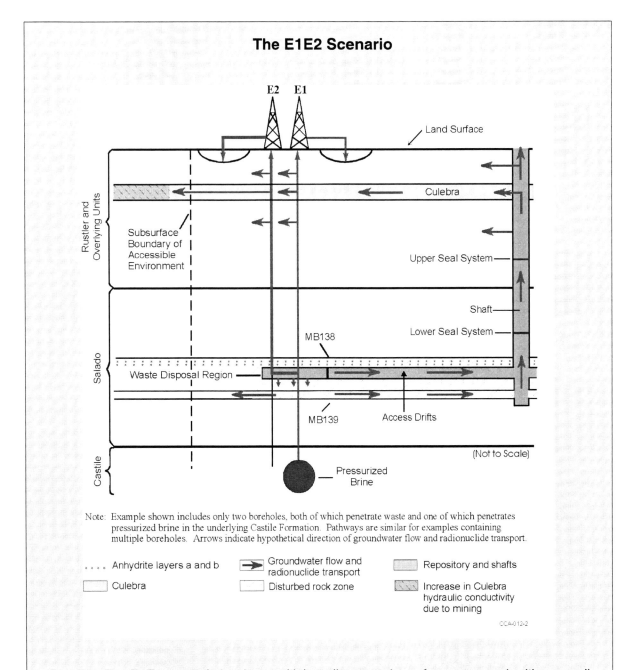

Box B.4: The E1E2 scenario involves multiple well penetrations of a waste panel, with one well penetrating a high-pressure brine panel below. Brine flows from a brine source through well E1 through the repository and is released through well E2. This flow path has the potential to bring large quantities of brine in direct contact with waste in the panel and to bring the contaminated brine to the overlying Salado or Culebra.

Appendix C

Biographical Sketches of Committee Members

Garrick, B. John, *Chair,* independent consultant, was a co-founder of PLG, Inc., an international engineering, applied science, and management consulting firm formerly in Newport Beach, California. He retired as president and chief executive officer in 1997. He received his B.S. degree in physics from Brigham Young University and his M.S. and Ph.D. degrees in engineering and applied science from the University of California, Los Angeles, and is a graduate of the Oak Ridge School of Reactor Technology. His professional interests involve risk assessment in fields such as nuclear energy, space and defense, chemical and petroleum and transportation. He is a past president of the Society for Risk Analysis. Dr. Garrick is a fellow of three professional societies and has received numerous awards, including the Society for Risk Analysis' Distinguished Achievement Award. He was appointed to the U.S. Nuclear Regulatory Commission's Advisory Committee on Nuclear Waste in 1994, of which he is now chairman. Dr. Garrick was elected to the National Academy of Engineering in 1993. He has been a member of the committee on the Waste Isolation Pilot Plant since 1989.

Mark D. Abkowitz, professor of civil engineering at Vanderbilt University and director of the Center for Environmental Management Studies, has an extensive background in environmental risk management, use of advanced information technologies in crisis management, and strategic and operational deployment of intelligent transportation systems. Dr. Abkowitz has been involved in hazardous materials transport education, research, product development, and technology transfer for manufacturers, transporters, regulators, and emergency response personnel. He has authored more than 70 journal publications and study reports, covering issues such as the risks of transporting high-level radioactive waste. Dr. Abkowitz has served on several national and international technical and advisory committees, including as chairman of the NRC Transportation Research Board standing committee on hazardous materials transport.

Alfred W. Grella, independent nuclear and hazardous materials transportation consultant, retired in 1990 from a career in U.S. government service, first at the DOT and later at the USNRC. His distinguished career spans 40 years as a professional in health physics, health protection, transportation, inspection and enforcement, training, and related regulatory activities. Mr. Grella received a bachelor's degree in chemistry from the University of Connecticut and completed the one-year management program at the National Defense University Industrial College of the Armed Forces. He has authored more than 30 published papers. He is a member of the American Nuclear Society and a fellow of the Health Physics Society. In 1965, the American Board of Health Physics awarded Mr. Grella certification as a health physicist. Mr. Grella received the M. Sacid (Sarge) Ozker Award in 1996 for distinguished service and eminent achievement in the field of radioactive waste management.

Michael P. Hardy, president of Agapito Associates, Inc., has experience in characterization, numerical modeling, design, and field experimentation for underground mines and high-level nuclear waste repositories at the BWIP site near Hanford, Washington, and Yucca Mountain, Nevada. Dr. Hardy is a member of the Society of Mining, Metallurgical and Exploration Engineers, Inc., and the American Society of Civil Engineers (ASCE). He is a former chairman of the Underground Technical Research Council, a joint ASCE-American Institute of Mining, Metallurgical, and Petroleum Engineers committee. Dr. Hardy received his bachelor of engineering from the University of Adelaide in Australia and his Ph.D. from the University of Minnesota in geoengineering.

Stanley Kaplan is one of the early practitioners of the discipline now known as quantitative risk assessment and a major contributor to its theory, language, philosophy, and methodology. Dr. Kaplan is a fellow of the Society for Risk Analysis and author of a number of seminal papers in this field. He is one of the first American contributors to the Russian science TRIZ, the Theory of the Solution of Inventive Problems, and currently consults and teaches in this area. He is a founder and board chairman of Bayesian Systems, Inc., a Washington-based company developing diagnostic, decision, simulation, and business management software. Dr. Kaplan is the recipient of several awards and honors, including the Society for Risk Analysis' Distinguished Achievement Award in 1996. Dr. Kaplan was elected to the National Academy of Engineering in 1999.

Howard M. Kingston, professor of chemistry and director of the Duquesne Environmental Research and Education Center at Duquesne University, has expertise in analytical chemistry techniques in environmental applications of hazardous waste characterization and remediation. His research interests include the development, automation, and standard encapsulation and transfer of analytical analysis methods. For the past several years, he has been actively involved in advancing the area of microwave sample preparation through basic research and the development of procedures that have been adopted by the EPA as standard methods. He has received numerous awards for his pioneering work and holds multiple patents in the fields of speciation, microwave chemistry, and chelation chromatography. He has co-edited and co-authored two of the American Chemical Society professional reference texts.

W. John Lee, Peterson Chair and professor of petroleum engineering at Texas A&M University, has expertise in petroleum reservoir analysis, pressure transient testing, and enhanced recovery methods. His professional experience includes research, operations, and consulting at Exxon Company U.S.A. and S.A. Holditch & Associates. He has received numerous awards from the Society of Petroleum Engineers, including the Reservoir Engineering Award, the Distinguished Service Award, the John F. Carll Award, Distinguished Membership, Distinguished Faculty Achievement Award, Distinguished Lecturer, and Honorary Membership. He is a member of the Georgia Tech Academy of Distinguished Engineering Alumni and he was elected to the National Academy of Engineering in 1993.

Milton Levenson, independent consultant, is a chemical engineer with more than 50 years of experience in nuclear energy and related fields. His technical experience includes work in nuclear safety, fuel cycle, water reactor technology, advanced reactor technology, remote control technology, and sodium reactor technology. His professional experience includes research and operations positions at the Oak Ridge National Laboratory, the Argonne National Laboratory, the Electric Power Research Institute, and Bechtel. Mr. Levenson is past president of the American Nuclear Society; a fellow of the American Nuclear Society and the American Institute of Chemical Engineers; and recipient of the American Institute of Chemical Engineers' Robert E. Wilson Award. He is the author of more than 150 publications and presentations and holds three U.S. patents. He received his B.Ch.E. from the University of Minnesota. He was elected to the National Academy of Engineering in 1976.

Werner F. Lutze, professor of chemical and nuclear engineering at the University of New Mexico (UNM) and director of the UNM Center for Radioactive Waste Management, has more than 25 years of research experience in materials science and geochemical issues relevant to the management of radioactive wastes, including selective mineral ion-exchange processes, repository near-field chemistry, waste form development, and trace analyses. He has published widely on weapons plutonium immobilization, waste disposal, and the chemistry of nuclear materials. Professor Lutze is a member of several professional organizations, including the Materials Research Society, the German Nuclear Society, and Sigma Xi.

Kimberly Ogden, associate professor of chemical and environmental engineering at the University of Arizona, has conducted research with Los Alamos National Laboratory collaborators to design treatment methods for remediating hazardous waste sites containing both toxic metals and organic materials, including plutonium-cellulose mixtures. She is also engaged in research investigating the merger of the semiconductors and biotechnology. Professor Ogden has authored or co-authored several book chapters, journal articles, and presentations. She is a member of several professional organizations including the American Institute of Chemical Engineers, the American Society of Engineering Education, and the American Chemical Society. She received her B.S.E degree in chemical engineering from the University of Pennsylvania and her M.S. and Ph.D. degrees from the University of Colorado.

Martha Scott, associate professor of oceanography at Texas A&M University, is a researcher in marine radiochemistry and geochemistry. Her present research involves radionuclide distribution in the Russian Arctic. Her work has dealt with the interaction between oceans and rivers, transport of materials in the marine environment, and chemistry of manganese nodules. The behavior of plutonium isotopes in rivers, estuaries, and marine sediments has been one of her longstanding research interests. She served for two years as an associate program director for chemical oceanography at the National Science Foundation (1992-1993). She received her Ph.D. degree from Rice University and was a National Science Foundation postdoctoral fellow at Yale University.

John M. Sharp, Jr., Chevron Centennial Professor of Geology, leads an active program in hydrogeology at The University of Texas at Austin. Professor Sharp has authored and co-authored more than 250 journal articles, books, reports, and presentations. His current research interests include characterization of groundwater flow and transport in fractured and karstic rocks; thermohaline free convection, hydrogeology of semi-arid zones, subsidence, and the effects of man on groundwater systems. He is a fellow of the Geological Society of America and recipient of its O.E. Meinzer Award (1979) and the American Institute of Hydrology's C.V. Theis Award (1996). Dr. Sharp is currently the editor of *Environmental and Engineering Geoscience* and the 2000 AT&T Industrial Ecology Fellow. He received his bachelor of geological engineering with distinction from the University of Minnesota and his M.S. and Ph.D. degrees in geology from the University of Illinois.

Paul G. Shewmon, emeritus professor of materials science and engineering at the Ohio State University, received a B.S. degree in metallurgical engineering from the University of Illinois and M.S. and Ph.D. degrees, also in metallurgical engineering, from the Carnegie Institute of Technology. He has lead work on fast breeder reactor materials at Argonne National Laboratory and served for 16 years on the USNRC's Advisory Committee on Reactor Safeguards. He has published 130 technical papers and two textbooks generally in the area of physical metallurgy and has received numerous awards for his research. He was elected to the National Academy of Engineering in 1979.

James E. Watson, Jr., professor of environmental sciences and engineering and director of the air, radiation, and industrial hygiene program at the University of North Carolina at Chapel Hill, holds a M.S. degree in physics from North Carolina State University and a Ph.D. in environmental sciences and engineering from the University of North Carolina at Chapel Hill. Professor Watson is accomplished in the fields of environmental radioactivity and radioactive waste management. He has received several awards for excellence in teaching, research, and service. He is a past president of the Health Physics Society and a past chairman of the Radiological Health Section of the American Public Health Association. He has served on the EPA's Radiation Advisory Committee and Executive Committee of the Agency's Science Advisory Board. He is a past chairman of the North Carolina Radiation Protection Commission and currently chairs the commission's Committee on Low-Level Radioactive Waste Management.

Ching H. Yew, an independent consultant and emeritus Halliburton professor of engineering mechanics at the University of Texas at Austin, received a B.S. degree in mechanical engineering from the National Taiwan University and M.S. and Ph.D. degrees in mechanical engineering from Cornell University and the University of California, Berkeley. Dr. Yew has specialized in solid mechanics and experimental mechanics, is a fellow of the American Society of Mechanical Engineers, and is a member of the Society of Petroleum Engineers. Dr. Yew has authored a book on the mechanics of hydraulic fracturing and published many articles concerning hydraulic fracturing and borehole stability.

Appendix D

Glossary

Actinide: Element with atomic number 90 (thorium) or greater.

Anhydrite: Anhydrous calcium sulfate.

Backfill: Earth or other material used to replace material removed during construction or mining. Backfill in excavations may or may not be the material originally removed. In the WIPP, magnesium oxide is the engineered backfill that replaces the mined salt and is placed in the free spaces surrounding the waste containers. Magnesium oxide is intended to chemically stabilize the radionuclides and minimize their solubility.

Borehole: Deep, circular hole of small diameter, such as an oil well or a water well.

Borehole Plugs: Engineered plugs to block the flow of liquid in either direction and to curtail the potential for movement of contaminants to the human environment. Several unplugged boreholes, presently being used to collect information for the WIPP, exist within the WIPP Land Withdrawal Area.

Brine: Water with dissolved salts at levels higher than seawater. Generally, brines are considered to have a total dissolved solids content of more than 100,000 milligrams per liter.

Brine Reservoir: Groundwater containing high levels of dissolved solids (brine) that may occur beneath the WIPP site either as discrete pockets (brine pockets) or as a saturated continuum. The committee uses the term "brine reservoir" to refer to both of these occurrences. At present, there is a great deal of uncertainty as to the location and form (i.e., discrete pocket or saturated continuum) of brine reservoirs beneath the WIPP repository.

Brucite: Magnesium dihydroxide, $Mg(OH)_2$.

Casing: Heavy metal pipe lowered into a borehole during or after drilling and cemented into place. It prevents the sides of the hole from caving and, prevents loss of drilling mud or other fluids into the hole.

Castile Formation: Oldest of the late Permian stratigraphic sequence of rocks, consisting of alternating layers of anhydrite and thin layers of limestone, with several thick layers of halite. See Figure 1.3.

Culebra Dolomite: Second-oldest member of the Rustler Formation ranging from approximately 7-8 meters thick at the WIPP site. The Culebra consists of dolomite with some clay minerals. Because it is a relatively transmissive unit, the Culebra is important to the groundwater flow model for the WIPP site.

Curie: Measure of the quantity of radioactive material in a sample, equal to 3.7×10^{10} disintegrations per second.

Cuttings: Rock chips cut by a bit in the process of well drilling and removed from the hole in the drilling mud in rotary drilling or by the bailer in cable-tool drilling. Well cuttings collected at closely spaced intervals provide a record of the strata penetrated.

Delaware Basin: Sedimentary basin in which the WIPP site is located. The Delaware Basin formed in the Permian sea and was gradually filled with thick, extensive layers of sediments and evaporite deposits.

Disturbed Rock Zone (DRZ): Zone around an excavation, in the host rock salt, where the stress field has been modified sufficiently to cause the formation of microfractures in the rock salt. Compared to the intact rock salt, the DRZ will have increased porosity because of the microfracturing, increased permeability as a result of interconnection of the microfractures, and decreased load-bearing capacity or strength.

Dolomite: Sedimentary rock consisting mostly of the mineral dolomite, calcium magnesium carbonate.

Dose: Energy imparted to matter in a volume element by ionizing radiation, divided by the mass of irradiated material in that volume element. The International System (IS) derived unit of absorbed dose is the gray (Gy); 1 Gy = 100 rad = 1 (Joule) per kilogram.

Drillbit or drill: A tool that cuts with its end by revolving or by a succession of blows.

Engineered Barriers: Man-made waste-isolating features that complement and strengthen natural waste-isolating barriers. These barriers are shaft seals, panel closures, borehole plugs, and backfill.

G-Value: Radiolytic yield unit. It corresponds to the number of molecules produced per 100 electronvolts of energy absorbed in the medium interacting with the ionizing radiation.

Half-Life: Time required for half of the atoms of a radioactive substance present at the beginning to disintegrate.

Hydraulic Fracture: Fracture of a rock in an oil or gas reservoir by pumping in water (or other fluid) and sand under high pressure. The purpose is to produce artificial openings in the rock to increase permeability. The pressure opens cracks and bedding planes, and sand introduced into these serves to keep them open when the pressure is reduced.

Hydrogen Getter: Material capable of capturing hydrogen gas.

Hydromagnesite: Mixed compound of magnesium carbonate and hydroxide, $4MgCO_3 \cdot Mg(OH)_2 \cdot 4H_2O$.

Injection Well: Well in an oil or gas field through which water, gas, steam, or chemicals are pumped into a reservoir or formation for pressure maintenance or secondary recovery, or for storage or disposal of the injected fluid.

Karst: Type of topography that is formed of limestone, gypsum, and other rocks by dissolution and is characterized by sinkholes, caves, and underground drainage.

Lithostatic Pressure: Pressure exerted by a column of overlying rock at a point in the earth's crust.

Magnesite: Magnesium carbonate, $MgCO_3$.

Marker Bed: Horizontally extensive formation that can be identified readily at different locations. For instance, the nonhalite interbed in the Salado, is numbered from the top of the Salado to the bottom and used to keep the repository at the same level within the Salado.

Panel Closures: Panel closures will limit the interaction of brine and gases among waste disposal panels. These closures will consist of a rigid concrete barrier and an isolation wall made of concrete construction block with an isolation zone between them.

Parameter: Algebraic symbol representative of a well-defined physical quantity with a numerical value. An adjustable parameter is envisioned to assume any value within its range (between the maximum and minimum numerical bounds). Any particular choice of a value renders a parameter a numerical constant.

Performance Assessment: Risk-based assessment of the safety performance of a nuclear waste facility.

Permeability: Capacity of a material to transmit fluids. A measure of the relative ease with which a porous medium can transmit a liquid under a potential gradient. Permeability depends on the size, shape, and degree of interconnectedness of pores and is generally measured in square meters. It is a property of the medium alone and is independent of the nature of the liquid.

pH: Measure of the acidity of a solution phase; negative logarithm of the hydrogen ion concentration.

Post-closure Period: Period beginning when the shafts of the disposal system are backfilled and sealed and ending 100 years later.

Pre-closure Period: Period between the beginning of operation and the time at which the shafts of the disposal system are backfilled and sealed. The operation period has been set as 35 years.

Radiogenic: Said of a product of a radioactive process.

Radiolysis: Decomposition brought about by ionizing radiation.

Radionuclide: Radioactive atom characterized by its mass and atomic number.

Retardation: Parameter that describes the ratio of the net apparent velocity of the concentration of a particular chemical species to the velocity of a non-reactive species. It is proportional to the slope of a sorption isotherm; thus, if the isotherm is nonlinear, the retardation factor is not constant and depends on solute concentration.

Rustler Formation: Second-youngest Permian Ochoan Formation, overlying the Salado, and consisting of five sequences (members) of thin-bedded strata. The lowermost beds consist of mudstone and sandstone interbedded with evaporites. The upper part of the formation consists of alternating evaporite and dolomite beds. The Culebra Dolomite member is the second member from the bottom of the formation. The total thickness of the Rustler Formation near the WIPP site is approximately 100 meters.

Salado Formation: Second-oldest Ochoan geologic formation consisting of a 230 million-year-old deposit of rock salt (halite) in near-horizontal beds; its total thickness lies between 200 and 400 meters. Very thin layers of clay, anhydrite, and potash minerals are interbedded with the halite beds. Lying at a depth of approximately 660 meters (2,160 feet) at the WIPP site, the Salado hosts the WIPP repository.

Salt Creep: Slow movement of salt over time as shear stresses cause movement within or between individual crystals. Mined salt "heals" as the creep restores its integrity.

Shaft: Vertical or inclined excavation through which a mine is worked.

Shaft Seals: Engineered barrier designed to limit fluid flow through the repository shafts. Once the repository has been filled, the entire column of each shaft will be backfilled with materials that prevent vertical flow of fluid. Materials include concrete, clay, asphalt, compacted salt, grout, and earthen fill.

Spallings: Chipping, fracturing, or fragmentation, and the upward and outward heaving, of rock caused by the interaction of a shock (compressional) wave at a free surface. Spallings in the WIPP can be caused by oil extraction and other human intrusion activities.

Transuranic (TRU) Waste: Radioactive waste consisting of radionuclides with atomic numbers greater than 92 in excess of agreed limits. A more precise definition, in DOE Order 5820.2A, EPA regulation 40 CFR 191, and the Land Withdrawal Act, is waste that is not high-level waste but is "contaminated with alpha-emitting radionuclides of atomic number greater than 92 and half-lives greater than 20 years in concentrations greater than 100 nanocuries per gram." The regulatory definition excludes actinide elements with atomic numbers between 90 and 92 (most significantly, thorium and uranium isotopes), which is in agreement with the literal meaning of "transuranic." However, common usage of the term "transuranic waste" is often understood to include all actinides.

TRUPACT-II: Transuranic Package Transporter, Model II. Container for road transport of contact-handled transuranic waste (see Figure 3.2). The TRUPACT-II container has been certified by the U.S. Nuclear Regulatory Commission.

Waste Characterization: Process of identifying and classifying the chemical, physical, and radiological constituents of each drum of waste.

Water Flooding: Technique used in the secondary recovery of petroleum and gas whereby water is injected into a petroleum or gas reservoir so that the pressure of the water expels the petroleum or gas.

Wattage Limit: In this report, the allowed maximum amount of heat generated by radioactive decay during transportation of TRU waste. The wattage limit for TRUPACT-II containers is 40 watts (40 joules per second).

Appendix E

Acronyms and Symbols

ASCE: American Society of Civil Engineers
CCA: compliance certification application
CCDF: Complementary Cumulative Distribution Function
CEMRC: Carlsbad Environmental Monitoring and Research Center
CH: Contact Handled
CH_4: Methane
CO_2: Carbon dioxide
DOE: U.S. Department of Energy
DOT: U.S. Department of Transportation
DRZ: Disturbed rock zone
EEG: State of New Mexico Environmental Evaluation Group
EPA: U.S. Environmental Protection Agency
H_2: Hydrogen
H_2S: Hydrogen sulfide
INEEL: Idaho Engineering and Environmental Laboratory
ITS: Intelligent transportation system
LWA: Land Withdrawal Act
MgO: Magnesium oxide
MTMH: Metric tons of heavy metal
N_2: Nitrogen
NORM: Naturally occurring radioactive material
NRC: National Research Council
PA: Performance assessment
RCRA: Resource Conservation and Recovery Act
RH: Remote handled
SNL: Sandia National Laboratories
TRANSCOM: Transportation Tracking and Communication

TRIZ: Theory of the Solution of Inventive Problems
TRU: Transuranic
TRUPACT-II: Transuranic Package Transporter, Model II
UNM: University of New Mexico
USNRC: U.S. Nuclear Regulatory Commission
WIPP: Waste Isolation Pilot Plant

Appendix F

Other Relevant Documents

Beauheim, R., G. Ruskauff, 1998. Analysis of Hydraulic Tests of the Culebra and Magenta Dolomites and Dewey Lake Redbeds Conducted at the WIPP Site. SAND98-0049. Albuquerque, NM: Sandia National Laboratories.

Beauheim, R. L, S. M. Howarth, P. Vaughn, S. W. Webb, and K. W. Larson, 1994. Integrated Modeling and Experimental Programs to Predict Brine and Gas Flow at the Waste Isolation Pilot Plant. GEOVAL '94 Validation Through Model Testing, OECD Documents—Safety Assessment of Radioactive Waste Repositories, Proceeding of an NEA/SKI Symposium, October 11-14. Paris, France. SAND 94-05996. Albuquerque, NM: Sandia National Laboratories.

Beauheim, R. L., W. R. Wawersik, and R. M. Roberts, 1993. Coupled Permeability and Hydrofracture Tests to Assess the Waste-Containment Properties of Fractured Anhydrite. International Journal of Rock Mechanics and Mining Sciences & Geomechanics Abstracts. 30(7):1159-1163.

Bodenstein, S., R. Gonzales, D. Sweetin, D. Taggart, S. Betts, J. Vigil, E. Derr, D. Yeamans, B. Sinkule, P. Rogers, and J. Harper, 1996. Transuranic Waste Characterization and Experimental Support at LANL Waste Characterization, Reduction and Repackaging Facility. Los Alamos, NM: Los Alamos National Laboratory. 483-484.

Bredehoeft, J., 1998. Drilling with Mud and Air into WIPP—Revisited (prepared for New Mexico Attorney General).

Bredehoeft, J., 1997. Rebuttal Technical Review of the Hartman Scenario: Implications for WIPP (Bredehoeft, 1997) by Swift, Stoelzel, Beauheim, Vaughn, and Larson. June 13, 1997, memorandum to EPA Compliance Certification Docket No. A-93-02.

Butcher, B., 1997. Waste Isolation Pilot Plant Disposal Room Model, SAND97-0794. Albuquerque, NM: Sandia National Laboratories.

Bynum, R., C. Stockman, H. Papenguth, Y. Wang, A. Peterson, J. Drumhansl, J. Nowak, J. Cotton, S. Patchet, M. Chu, 1998. Identification and Evaluation of Appropriate Backfills for the WIPP. SAND98-1026C. May. Albuquerque, NM: Sandia National Laboratories.

Callahan, G. D., 1999. Crushed Salt Constitutive Model. SAND98-2680. February. Albuquerque, NM: Sandia National Laboratories.

Channell, J. K., B. A. Walker, 2000. Evaluation of Risks and Waste Characterization Requirements for the Transuranic Waste Emplaced in WIPP During 1999. EEG-75. Albuquerque, NM: Environmental Evaluation Group.

Chaturvedi, L., J. Channell, 1985. The Rustler Formation as a Transport Medium for Contaminated Groundwater. EEG-32. Albuquerque, NM: Environmental Management Group.

Chaturvedi, L., T. Clemo, M. K. Silva, and W. W. L. Lee, 1997. Conceptual Models Difficulties in the WIPP Compliance Certification Application. In Proceedings of the Sixth International Conference on Radioactive Waste Management and Environmental Remediation, ICEM '97, October 12-16, 1997, Singapore. R. Baker, S. Slate, and G. Benda, eds. New York, NY: American Society of Mechanical Engineers. 423-427.

Clemo T., L. Chaturvedi, W. Lee, 1997. Problems with Data Used in the WIPP Certification Application Performance Assessment. In Proceedings of the Sixth International Conference on Radioactive Waste Management and Environmental Remediation, ICEM '97, October 12-16,1997. Singapore. R. Baker, S. Slate, and G. Benda, eds. New York, NY: American Society of Mechanical Engineers. 1009-1012.

Connolly, M., S. Kosiewicz, 1997. TRU Waste Transportation: The Flammable Gas Generation Problem. Technology: Journal of the Franklin Institute. 334(A):351-356.

Economy, K. M., J. C. Helton, and P. Vaughn, 1999. Sandia Brine and Gas Flow Patterns Between Excavated Areas and Disturbed Rock Zone in the 1996 Performance Assessment for the Waste Isolation Pilot Plant for a Single Drilling Intrusion that Penetrates Repository and Castile Brine Reservoir. SAND99-1043. October. Albuquerque, NM: Sandia National Laboratories.

Chaturvedi, L. and J. K. Channell, 1985. The Rustler Formation as a Transport Medium for Contaminated Groundwater. EEG-32. Albuquerque, NM: Environmental Evaluation Group.

Environmental Protection Agency, 1998. EPA's Analysis of Air Drilling at WIPP. Docket No: A-93-02 (IV-A-1) EPA. January 27, 1998. Washington, DC: Environmental Protection Agency.

Fanghänel, T., J. Kim, P. Paviet, R. Klenze, W. Hauser, 1994. Thermodynamics of Radioactive Trace Elements in Concentrated Electrolyte Solutions: Hydrolysis of Cm3+ in NaCl-Solutions. Radiochimica Acta. 66-67: 81-87. Germany.

Federal Register, 1990. Waste Analysis Plans and Treatment/Disposal Facility Testing Requirements. June 1, 1990. Federal Register. 55 (106):22669.

Federal Register, 1997. Joint NRC/EPA Guidance on Testing Requirements for Mixed Radioactive & Hazardous Waste. November 20, 1997. Federal Register.

Felmy, A., and D. Rai, 1992. An Aqueous Thermodynamic Model for a High Valence 4:2 Electrolyte Th4+ -SO_2^{-4} in the System Na^+ - K^+ - Li^+ - NH^{+4} - Th^{4+} - SO_4^{2-} – HSO^{-4} – H_2O to High Concentration. Journal of Solution Chemistry. 21(5): 407–423.

Francis, A. J., J. B. Gillow, and M. R. Giles, 1997. Microbial Gas Generation Under Expected WIPP Repository Conditions. SAND96-2582. March. Upton, NY: Sandia National Laboratories.

Gray, D. H., J. W. Kenney, S. C. Ballard, 2000. Operational Radiation Surveillance of the WIPP Project by EEG During 1999. EEG-79. September. Albuquerque, NM: Environmental Evaluation Group.

Greenfield, M. A. and T. J. Sargent, 2000. Probability of Failure of the Trudock Crane System at the Waste Isolation Pilot Plant. EEG-74. May. Albuquerque, NM: Environmental Evaluation Group.

Helton, J.C., et. al., 1998. Uncertainty and Sensitivity Analysis Results Obtained in the 1996 Performance Assessment for the Waste Isolation Pilot Plant, SAND98-0365. Albuquerque: NM: Sandia National Laboratories.

Idaho National Engineering Enviromental Laboratory (INEEL), 1998. TRUPACT-II Matrix Depletion Program Final Report. INEEL/EXT-98-00987, Rev. 0. September. Idaho Falls: INEEL.

Kenney, et al., 1999. Preoperational Radiation Surveillance of the WIPP Project by EEG From 1996-1998. EEG-73. October. Albuquerque, NM: Environmental Management Group.

Kersting, A. B., D. W. Efurd, D. L. Finnegan, D. J. Rokop, D. K. Smith, and J. L. Thompson, 1999. Migration of plutonium in ground water at the Nevada Test Site, Nature. January. 397(6714):56-59.

Knowles. M. K. and K. M. Economy, 2000. Evaluation of Brine Inflow at a Waste Isolation Pilot Plant (abstract). Water Environment Research. 72(4):397-404.

Krumhansl, J. L., M. A. Molecke, H. W. Papenguth, and L. H. Brush, 1999. Historical Review of Waste Isolation Pilot Plant Backfill Development. SAND99-0404A. Albuquerque, NM: Sandia National Laboratories.

Lechel, D. J. and C. D. Leigh, 1998. Plutonium-238 TRU Waste Decision Analysis. SAND98-2629. December 31, 1998. Carlsbad, NM: Sandia National Laboratories.

Lucero, D. A., G. O. Brown, and C. E. Heath, 1998. Laboratory Column Experiments for Radionuclide Adsorption Studies of the Culebra Dolomite Member of the Rustler Formation. SAND97-1763. April. Albuquerque, NM: Sandia National Laboratories.

Madic, C., 2000. Toward the End of PuO2's Supremacy, Science Magazine. 287:243-244.

Mellegard, K. D., T. W. Pfeifle, and F. D. Hansen, 1999. Laboratory Characterization of Mechanical and Permiability Properties of Dynamically Compacted Crushed Salt. SAND98-2046. Albuquerque, NM: Sandia National Laboratories.

Mercer, J. W., D. L. Cole, and R. S. Holt, 1998. Basic Data Report for Drillholes on the H-19 Hydropad. SAND98-0071. Albuquerque, NM: Sandia National Laboratories.

Molecke, M., 1979. Gas Generation from Transuranic Waste Degradation: Data Summary and Interpretation (SAND 79-1245). December 1979. Albuquerque, NM: Sandia National Laboratories.

APPENDIX F: OTHER RELEVANT DOCUMENTS

NEA/OCDE, 1999. Progress towards Geologic Disposal of Radioactive Waste: Where Do We Stand? Radioactive Waste Management. NEA/OCDE.

Neill, R., L. Chaturvedi, D. Rucker, M. Silva, B. Walker, J. Channell, T. Clemo, 1998. Evaluation of the WIPP Project's compliance with the EPA Radiation Protection Standards for Disposal of TRU Waste; EEG-68, Albuquerque, NM.

Oversby, V. M., 2000. Plutonium Chemistry Under Conditions Relevant for WIPP Performance Assessment–Review of Experimental Results and Recommendations for Future Work. EEG-77. September. Albuquerque, NM: Environmental Evaluation Group.

Perkins, G. W., and D. A. Lucero, 1998. Interpretation of Data Obtained from Non-Destructive and Destructive Post-Test Analyses of an Intact-Core Column of Culebra Dolomite. SAND98-0878. Albuquerque, NM: Sandia National Laboratories.

Perkins, W.G., D.A. Lucero, and G.O. Brown, 1998. Column Experiments for Radionuclide Adsorption Studies of the Culebra Dolomite: Retardation Parameter Estimation for Non-Eluted Actinide Species. SAND98-1005. Albuquerque, NM: Sandia National Laboratories.

Pfeifle, T. W. and F. D. Hansen, 1998. Database of Mechanical and Hydrological Properties of WIPP Anhydrite Derived from Laboratory-Scale Experiments. SAND98-1714. Albuquerque, NM: Sandia National Laboratories.

Rechard, R. P., 1998. Milestones for Disposal of Radioactive Waste at the Waste Isolation Pilot Plant in the United States. SAND98-0072. April. Albuquerque, NM: Sandia National Laboratories.

Rocker D. F., 1998. Sensitivity Analysis of Performance Parameters Used in Modeling the WIPP – Summary. Dale F. Rucker, ed. EEG-69. May. Albuquerque, NM: Environmental Evaluation Group.

Rucker, D. F., 2000. Probabilistic Safety Assessment of Operational Accidents at the Waste Isolation Pilot Plant. EEG-78. September. Albuquerque, NM: Environmental Evaluation Group.

Silva, M., 1996. Fluid Injection for salt Water Disposal and Enhanced Oil Recovery as A potential Problem for the WIPP: Proceedings of a June 1995 Workshop and Analysis. EEG-62. Albuquerque, NM.

Sandia National Laboratories, 1979. Summary of Research and Development Activities in Support of Waste Acceptance Criteria for WIPP. SAND79-1305. November. Albuquerque, NM: Sandia National Laboratories.

Sandia National Laboratories, 1997. Spalling Model Position Paper. Semi-Analytical Calculations Conducted in Support of Alternative Spallings Model. SECF-A:2.01.5.3.1. WPO# 43214. Albuquerque, NM: Sandia National Laboratories.

Sandia National Laboratories, 1999. Statistical Analyses of Scatterplots to Identify Important Factors in Large-Scale Simulations, SAND98-2202. Albuquerque, NM: Sandia National Laboratories.

Stoelzel D. M. and, P. Swift, 1997. Technical review comment resolution for sensitivity of the length of fracture approximations in BRAGFLO to the grid used in *Supplementary Analyses of the Effect of Salt Water Disposal and Waterflooding on the WIPP*. WPO #44158. June.

Telander, M. R., and R. E. Westerman, 1997. Hydrogen Generation by Metal Corrosion in Simulated WIPP Environments. SAND96-2538. Albuquerque, NM: Sandia National Laboratories.

U.S. Department of Energy (DOE), 1994. Report on the Emergency Response Training and Equipment Activities through Fiscal Year 1993 for the Transportation of Transuranic Waste to the Waste Isolation Pilot Plant. Revision to DOE/WIPP 92-055 (November 1992). DOE/WIPP 93-061. April. Albuquerque, NM: DOE.

U.S. Department of Energy, 1995. Emergency Planning, Response, and Recovery Roles and Responsibilities for TRU-Waste Transportation Incidents. DOE/CAO-94-1039. January. Nevada: DOE Albuquerque Operations Office and Carlsbad Area Office.

U.S. Department of Energy, 1996. Transuranic Waste Characterization Quality Assurance Program Plan: Interim Change. CAO-94-1010. November 16. Carlsbad, NM: U.S. Department of Energy, Carlsbad Area Office.

U.S. Department of Energy, 1998. TRU Waste Characterization Quality Assurance Program Plan. CAO-94-1010 Revision 1.0. December 18. Carlsbad, NM: U.S. Department of Energy, Carlsbad Area Office.

U.S. Department of Energy, 2000. Waste Isolation Pilot Plant: Pioneering Nuclear Waste Disposal. USDOE/CAO-00-3124. February. Carlsbad, NM: U.S. Department of Energy, Carlsbad Area Office.

U.S. Department of Energy, 2000. CH-TRU Waste Transportation System. Rail Study. DOE/WIPP 00-2016. September. Albuquerque, NM: U.S. Department of Energy, Carlsbad Area Office.

Westinghouse Electric Co., 1997. TRUPACT-II Payload Expansion Plan. December. Carlsbad, NM: Westinghouse Electric Co., Waste Isolation Division.

WIPP, 1996. Lower-tier Monitoring Plan: Groundwater Surveillance Program Plan, WP 02-1, Revision 3. Online. Available at www.wipp.carlsbad.nm.us/wipp.htm.

WIPP, 1997. Lower-tier Monitoring Plan: Delaware Basin Drilling Surveillance Plan, WP 02-PC.02, Revision 0. Online. Available at www.wipp.carlsbad.nm.us/wipp.htm.

WIPP, 1997. Lower-tier Monitoring Plan: WIPP Waste Information System Program, WP 05-WA.02, Revision 0. April 15. Online. Available at www.wipp.carlsbad.nm.us/wipp.htm.

WIPP, 1998. Lower-tier Monitoring Plan: WIPP Geotechnical Engineering Program Plan, WP 07-01, Revision 2. Online. Available at www.wipp.carlsbad.nm.us/wipp.htm.

WIPP, 1997Lower-tier Monitoring Plan: WIPP Underground and Surface Surveying Program, WP09-ES.01, Revision 1. Online. Available at www.wipp.carlsbad.nm.us/wipp.htm.